Reducing Risk

Participatory learning activities for disaster mitigation in Southern Africa

Astrid von Kotze and Ailsa Holloway

Distributed internationally by Oxfam United Kingdom and Ireland (UK/I) Publishing (a member of Oxfam International) as a contribution to dissemination of current development practice. Oxfam UK/I does not necessarily endorse methodologies, opinions or views expressed in this publication.

Distributed by agents for Oxfam UK/I in the following territories:
In the USA and Canada: Humanities Press International, 165 First Avenue, Atlantic Highlands, NJ 07716-1289
In Southern Africa: David Philip Publishers, PO Box 23408, Claremont, Cape Town, 7735, South Africa
Throughout the world: Oxfam UK and Ireland, 274 Banbury Road, Oxford OX2 7DZ, United Kingdom

Layout and cover design: Lesley Lewis, Inkspots, Durban, Natal
Cover photograph: Guy Tillim
Editor: Sonya Keyser
Printer: Kohler, Carton and Print, Durban, Natal

The photographs in this book were taken during the training sessions of the Southern Africa Disaster Management Training Programme. They show participants from the ten Southern African countries who are acknowledged on page vii.

ISBN: 0 85598 347 7

The International Federation of Red Cross and Red Crescent Societies promotes the humanitarian activities of National Societies among vulnerable people. By coordinating international relief and encouraging development support it seeks to prevent and alleviate human suffering and so contribute to peace in the world. The Federation, the National Societies and the International Committee of the Red Cross together constitute the International Red Cross and Red Crescent Movement.

Contents

Section 1: Key Concepts and Risk Factors

key concepts

Section 2: Risk Assessment: Community-based Considerations

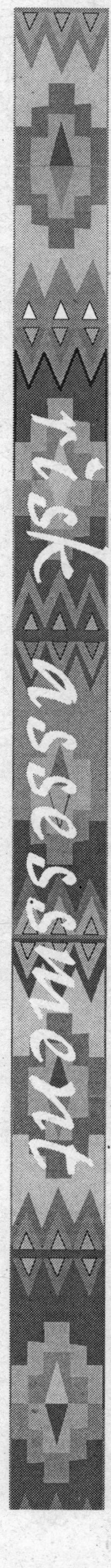

Section 3: Risk Reduction and Emergency Preparedness and Planning

risk reduction

risk reduction

Foreword

As this book nears completion, storm clouds loom over much of Southern Africa. They bring with them the promise of life-giving rain — to fill our thirsty reservoirs, water this year's harvest, and to cool parched, dry throats.

Southern Africans have once more withstood the hardships induced by a severe drought, just as they have managed seasonal floods, and outbreaks of dysentery, as well as decades of civil unrest. Today, in the region's rural areas, in isolated villages and in peri-urban settlements, at-risk households and communities continue — somehow — to find ways of managing mounting socio-economic and environmental pressures that increase their vulnerability to both natural and other threats.

"*Reducing Risk: Participatory Learning Activities for Disaster Mitigation in Southern Africa*" is one tool for those who work with at-risk communities, either in development or relief. It is a series of participatory learning activities intended to increase understanding about community risk and vulnerability, and strengthen the training capacities of those involved in community-based disaster management.

"*Reducing Risk*" is based on four assumptions. First, it assumes that risk reduction efforts are more effective and sustainable if they are linked with or integrated into existing community-based services. Second, it assumes that the relationships between agencies involved in disaster mitigation and communities at-risk should be those of active partnership. Third, it assumes that emergency operations are opportunities for promoting prevention, mitigation, preparedness and recovery — as well as relief. Last, because in Southern Africa, the greatest impact of recurrent threats falls on women, it assumes that disaster-related initiatives must consider gender, as well as actively involve women in their design and implementation.

"*Reducing Risk*:" is not an all-inclusive publication which precisely describes methods for assessing hazards and vulnerabilities. Nor is it a definitive text on disaster management. It is not a book that details risk-reduction measures related to cholera management, drought-resistant seed production, or other fields requiring specialist input.

However, it is hoped that this collection of activities designed for adult learners will better equip development and relief workers with techniques that maximise participant learning related to risk, and its reduction at community level.

To date, there have been no learning and teaching materials published in Southern Africa, that support development and relief workers who consider community-based disaster mitigation a priority. This book is a first step in bringing the concept of risk assessment and reduction alive in Southern Africa through a participatory learning process.

Acknowledgements

Reducing Risk grew out of a disaster management training programme sponsored by the International Federation of the Red Cross and Red Crescent Societies in southern Africa. Its creation and production were coordinated by the Federation's regional delegation in southern Africa, based in Harare, Zimbabwe, and the Department of Adult and Community Education, University of Natal, in Durban, South Africa. Many people provided invaluable support and input during the course of the *Southern African Disaster Management Training Programme*.The publishers would specifically like to express its appreciation to those listed below:

Donors

Canadian International Development Agency (CIDA), the British, Canadian and Japanese Red Cross Societies, and the British Government's Overseas Development Administration

Programme Coordinator

Diane Lindsey

Participants in the Southern African Disaster Management Training Programme

Jane Chagie, Paulina Chiziane, Harold Dzumani, Mandisa Kalako-Williams, Oreneile Kristensen, Timothy Mbewe, Alfred Mhone, Dama Mosweunyane, Justin Mukwecheni, Dorothy Mwimanenwa, Emmanuel Ndlangamandla, Neo Ramarou, Malefelisa Setloboko, Ann Solomon
and
Alfredo Chissano, Ephraim Matinhira, Eunice Mucache, Prince Morare, Rui Possolo, Marson Sharpley, Fernanda Teixeira

Technical Resource People

Mary Basset, Phoebe Mbembya, Graham Eele, Gary Eilerts, Stefan Helming, Terry Jeggle, Ruud Jensen, Kamal Laldas, Freda Luhila, Elisabeth Mason, Lucy Muyoyeta, Juan Saenz, Pauline Stanford, Carol Thompson

Editorial Committee — *Reducing Risk*

Heather Cameron, Munyaradzi Chenje, Graham Eele, Mandisa Kalako-Williams,

Regional Programme Advisors

Roger Buckland, Denis Nkala, JacquiePike,
K. Pushpanath, Margaret Samuriwo, John Undulu

Logistical and Administrative Support

Moline George, Zanele Mabaso, Tauya Machakaire, Forshow Magorongah

Field Training

Staff and volunteers of Mozambique Red Cross, and residents of Ndzadzo Village,
Tete, Mozambique

Thanks, also to colleagues all over the world who generously gave of their time for discussions and who provided us with valuable materials.

About the Authors

Astrid von Kotze is a programme coordinator in the Department of Adult and Community Education of the University of Natal, South Africa. She is the author of two books on community and worker education. She has extensive experience in designing and facilitating participatory planning and learning processes. Her previous work includes creative writing, story-telling and play-making programmes for working people and youth, and contributing to the drafting of adult education policy in South Africa. Astrid has a doctorate in Literary Studies from the University of the Witwatersrand, Johannesburg.

Since March 1992 **Ailsa Holloway** has been working for the International Federation of Red Cross and Red Crescent Societies as the regional disaster preparedness advisor for Southern Africa, Zimbabwe. Her previous work experience includes being health advisor in emergency operations for the UNHCR, consultant for the Global Disaster Preparedness Programme of the WHO, and editor of the IDNDR. Ailsa has a doctorate in Public Health from the University of California, Los Angeles.

Introduction

Who this book is for

* practitioners who are concerned about strengthening the capacities of vulnerable communities to deal with emergencies such as those caused by drought, epidemics, floods and population movements;
* development workers in the Southern African region, particularly those who work at the grassroots level;
* development workers who are looking for activities that are specific to Southern Africa's risk profile;
* people who are concerned with long-term sustainable development; i.e. who realise that development programmes need to incorporate risk reduction into their plans, as experience has shown that programmes get seriously disrupted by emergencies and disasters;
* people working in participatory training for community development and who would benefit from a more experiential approach to education and training.
* trainers who are working in the area of relief and development and who believe a participatory approach based on the development of understanding and skills will lead to more informed and effective risk reduction interventions;
* disaster management trainers who wish to re-focus from rehabilitation / reconstruction after emergencies to mitigation and risk reduction before disaster strikes;
* trainers who have reviewed disaster management training literature in search for a more applied approach and who are looking for learning modules rather than technical information;
* personnel working in the field of aid / relief, who have been confronted repeatedly with the same kinds of problems that arise when communities are at-risk, and who believe that relief action should be coupled with interventions aimed at development.

This book aims to make a beginning for those who may not have the confidence to attempt initiating participatory learning for disaster reduction and those who wish to try out a new approach to training.

It is not an information-based disaster managers' training manual; for materials that are more technically oriented please see the reference list at the end of this book.

How this book came about

In 1991-92 Southern Africa was severely affected by the region's "worst drought in living memory". The response which followed drew together the international community with Southern Africa's governments, NGOs and private companies to mitigate the drought's impact.

The demands of this regional effort were particularly taxing for Southern Africa's NGOs and National Red Cross Societies, who found themselves working as willing but often overextended operational partners. As such, the 1992-93 drought response highlighted at least three pressing priorities for NGOs working in disaster-prone communities. Firstly, it underlined the urgent need to strengthen the disaster management capacities of Southern Africa's NGOs even further. Secondly, it emphasised the importance of developing disaster management training materials and approaches that are specific to this region's risk profile. And thirdly, the 1992-93 relief response underscored the need for a disaster management training approach that was active and participatory — to facilitate the greatest learning.

In partnership with the Department of Adult and Community Education at the University of Natal, the Regional Delegation of the International Federation of Red Cross and Red Cross Crescent Societies designed a training project to address those needs. This project, which has lead to *Reducing Risk* began in 1993, after a process of consultation with Southern Africa's NGOs and National Red Cross Societies. Over the course of the next eighteen months, sixteen NGO and Red Cross representatives from ten Southern African countries completed a series of six one-week workshops in disaster management.

Reducing Risk is based on that six week course. Most of the learning activities in this book were developed in the following way:

They were designed for specific sessions; tried out during the course, facilitated by the writers and other experts (as acknowledged in the text), and redrafted.

They were evaluated in terms of both learning process, and learning outcomes.

The written activities were reviewed and discussed with course participants; in some instances they were again tested in class or by participants acting as facilitators in their own workshops.

In response to feed-back received they were edited and reworked / written.

They were reviewed and edited a number of times by an editorial advisory board comprising subject specialists, educators and SADMTP participants.

Finally, the laid-out draft of the book was reviewed and edited numerous times.

Guiding Principles for the Preparation of these Learning Activities

The main principle guiding the production of the learning activities is simply the belief that people learn best when they are respected and when they enjoy the learning process. Beyond that, what is particular to these activities?

(1) They are context-specific: The learning priorities addressed by these activities are intended to fit Southern Africa's specific risk profile. Both the choice of a learning process that emphasises oral communication and story-telling, and the examples given are rooted in the context of actual field situations from which participants come and to which the learning will be transferred.

(2) They are experience-based: Learning about disaster reduction should be based on participants' own experience of dealing with daily and recurrent risks. By sharing and comparing the kinds of strategies they have developed in order to cope with and address particular vulnerabilities, participants acknowledge each other as invaluable sources of knowledge. This experience-based learning approach aims to enhance learners' ability to solve problems in the field logically and independently, drawing on available resources and/or identifying additional sources of information and materials. Hence, many of these disaster management learning activities aim at developing participants' analytical thinking and creative problem-solving skills.

(3) They are participatory: In order to be effective disaster reduction should be a participatory process, involving members of affected communities at all stages of a programme. The activities in *Reducing Risk* encourage participants to monitor and manage their own learning processes and in this way the responsibility for learning is transferred from the facilitators to the participants.

Communication skills are an important way of enhancing learners' ability to fully participate in the training process. Learning activities encourage cooperative group work and interactive faculties are further developed through a focus on active listening skills, analysis of assumptions, the ability to formulate questions and various methods of reporting and presenting information.

(4) They are analytical: The process aims to develop learners' critical thinking, planning and response skills. Every at-risk community faces different pressures, and requires that decisions and actions taken are based on an analysis of the particular dynamics and pressures dictated by that scenario. Participants practise how to assess and analyse the causes and effects of different kinds of risks and how to apply a variety of creative problem-solving skills to particular hazards. *Reducing Risk* presents a shift from training based on technical information aimed at response to emergency situations, to helping participants learn how to plan for more long-term risk reduction.

(5) They are applied: The purpose of any disaster management training must be to develop the ability of participants and their organisations to respond more efficiently and effectively to threats posed by the impact of hazards on vulnerable communities. This means that learners should acquire practical skills and understanding that enable them to intervene usefully.

How to use this Book

This book is primarily a tool for Southern African practitioners working with highly vulnerable communities. If you are planning to facilitate a series of learning activities in disaster reduction you may want to use this book both as a reference and as a 'how to' manual.

If you work as a development agent *Reducing Risk* should provide you with ideas and suggestions for training / learning activities. If you are already experienced in facilitating and leading participatory learning activities you may want to use this book as a resource for activities with a particular focus on risk reduction in the Southern African context. Alternatively, you may need more detailed instructions on how to initiate, manage and monitor and evaluate learning activities, and in this case this book gives you detailed step-by-step guidelines of how to facilitate a session.

You do not have to begin with activity 1 in section 1 and go all the way through to the last activity in section IV. Please read through this book and choose those activities that would best meet your training needs. If you are in the process of constructing a whole course in disaster reduction you may find that this book provides you with a core curriculum.

We would strongly advise you to include two further components into a programme:

(1) a training workshop in Participatory Rural Appraisal (PRA), and

(2) activities that strengthen participants' ability to plan and develop programme proposals.

Reducing Risk is viewed as one tool for empowering Southern Africa's most at-risk communities to decrease their vulnerability to recurrent threats. The reference section at the back gives some useful information on other training materials.

Finding your way around this book

Notes for Facilitators (page xvi)

The Notes for Facilitators outline some of the essential components that need to be considered and incorportated when planning a learning process. This section has useful comments about the different roles facilitators take on in participatory learning processes, and outlines some of the activities that have to be part of how to get a training and learning process started. There are specific descriptions on facilitating various learning processes, and examples of energising games, exercises that improve listening skills and processes to maintain group learning and active participation.

Section 1: Key Concepts and Risk Factors (page 1)

This section contains ten learning activities that are mostly conceptual in nature, but practical in application. They aim at developing participants' understanding of concepts and terminology with regard to risk, both as it shapes the daily lives of people and communities, and as it impacts on the work of non-governmental and humanitarian organisations. The activities in this section 'set the stage' for further learning activities, as they stimulate critical thinking around issues of risk reduction as an essential component of development.

Section 2: Risk and Capacity Assessment: Community-based Considerations (page 93)

Section 2 comprises eight practical learning activities that aim at developing participants' understanding of different approaches to risk assessment in at-risk communities. Participants experience and practice a number of different processes for gathering and analysing data, including visual mapping, participatory rural appraisal and questionnaire administration. There are a host of techniques for assessing risk. These activities focus on those most suitable for use at community level. They aim to improve participants' learning skills, particularly those which assess risks and capacities at individual, household and community levels.

Section 3: Emergency Preparedness and Risk Reduction (page 187)

Section 3 is made up of twelve participatory learning activities that further improve participants' ability to incorporate risk reduction into emergency response plans. The activities are a mixture of analytical thinking and strategic planning exercises; participants practice how to identify appropriate strategies for reducing vulnerability, and how to examine proposed action plans in terms of gender considerations. This section aims at developing participants' planning skills by asking them to consider the impact of various hazards on elements at risk.

The Organisation of Individual Activities

Please note:

- ➠ It is **critically important** that you read through an entire learning activity first, and get an idea of what is required of you and the participants, and to ensure that you are clear about the purpose and procedure of the activity.
- ➠ Once you have decided to conduct an activity work out what you can and need to prepare in advance.
- ➠ Plan your evaluation in advance: how will you know whether the activity has achieved its purpose? What indicators of learning will you use?

Each activity is structured in the following way:

We begin by stating the **purpose**; this gives you a general idea of what participants are expected to learn.

The **note** pinned to the purpose further explains what the learning is about,and why and how we think it is important.

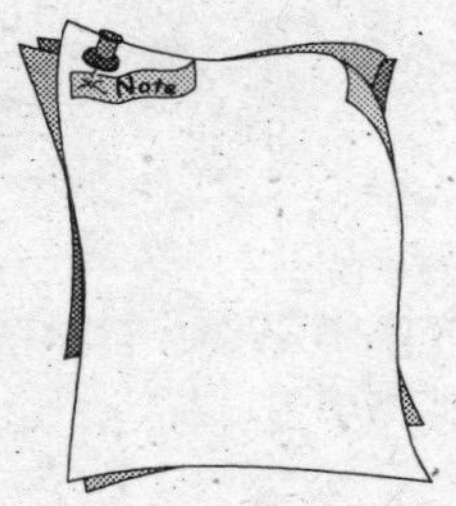

The **procedure** describes the type of learning activity entailed: it tells you whether this is a role play, or an input-based discussion, or a task based on a case scenario, etc.

The **time** states the approximate duration of the activity; please note that all the activities are based on a relatively small group of about 16 participants. If you are working with a larger group you need to increase the time allocated, particularly in processes involving discussion or report-backs.

Materials lists the resources and materials involved in the activity; when specific resources such as case scenarios or cards are needed these have been included in the resources.

Please note that we make the asssumption that flipchart, broad pens and sticky stuff — such as 'blue tack', prestik, or masking tape — are freely available.

Using flipcharts It is important to signal your respect for participants' experience and knowledge. Ask different participants to record suggestions and information on big sheets of paper or flipchart, and display the flipchart on walls throughout the duration of a particular focus. Allocate time during which such information can be copied, or write it up and distribute it to participants. This will be a useful record of the learning session, and a reference that can form the basis of report-backs to participants' organisations and agencies.

Process

This heading signals the beginning of a described activity.

Every process begins with the heading **Introduction**. This part of an activity serves to focus the attention and get participants thinking around the particular issue at hand. In some cases the introduction includes an input with background information that is needed before participants can do the ensuing task.

The first step of each introduction is '**outline the purpose and procedure of this activity**'. This instruction is very important and should be followed carefully. In order to be able to participate actively participants need to know what the purpose of an activity is, and what will be expected of them in terms of the procedure. When preparing to facilitate an activity you would do well to make notes that allow you to enact this first instruction. Such notes allow you to check on your understanding of the activity and 'rehearse' the explanation and description that you will give.

Participant action

This indicates the beginning of a task, involving primarily participants. Please look at the **facilitator's notes** for ideas on how to 'divide participants into groups'.

The **flipchart visual** is our way of showing an instruction given to participants. We have found that it helps to write such instructions on flipchart, or even write them on strips of paper and distribute them to participants.

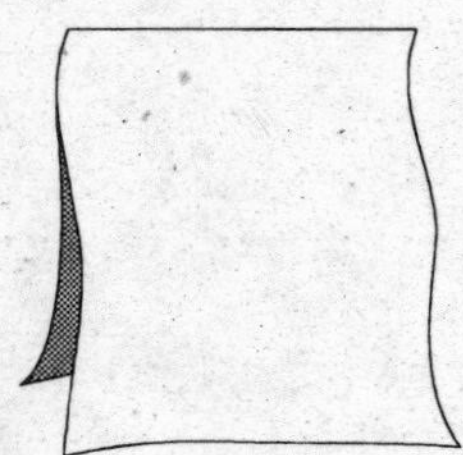

We have used a **shaded flipchart** to show examples of feedback, or something written up in response to a task or instruction given to participants.

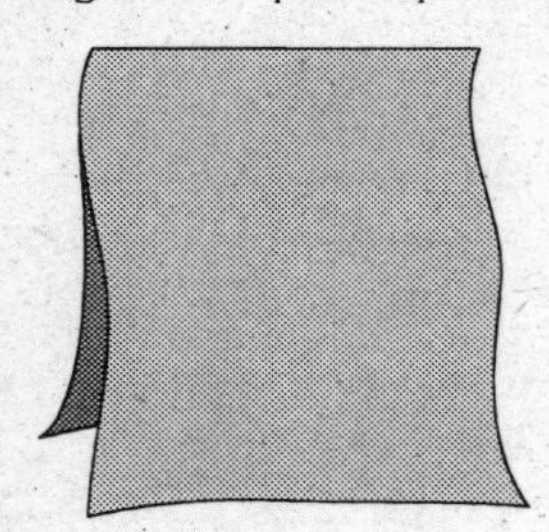

Brainstorming is a useful way of generating a lot of information very quickly. It is a particularly useful tool for creative problem-solving. Encourage participants to say whatever comes into their minds in response to a topic; explain that this stimulates creative thinking, and point out that in this process no idea is too far-fetched: anything goes. Ensure that all responses are recorded without 'censorship', and follow the brainstorm with sorting and processing of the information.

In a **buzz** two or three participants briefly exchange ideas on any given topic.

Role Plays need to be initiated and managed with care. Give clear instructions to as to the purpose and procedure of the role play. Ensure that participants are clear about what is needed from them, individually and collectively. Decide what information you have to give in advance, explain the difference between 'playing a role' and 'being a character', and take care in de-briefing players at the end of the play. Role plays often generate strong emotions and these have to be acknowledged and dealt with before participants are able to maximise their reflection and learning.

Monitor the process and assist where necessary — when participants have gone into groups (or settled to work individually) to do a given task you should go around to check whether they are clear about what to do. Ask them if they need an explanation, or encourage them to check their understanding of the task with you. Ask questions such as 'do you know what to do?' or 'do you need / want some help?' After that, participants often work best when left alone 'to get on with it'. Visit groups at a later stage to check on their progress and to remind them of the time limit. We have often found that groups do not plan their report-backs unless reminded to do so. However, as a general rule interrupt as little as possible. If the group spirit is right participants will call on you when they need your assistance!

The instruction **Initiate and facilitate the report-back** appeals to you in your role as process manager. This stage usually also necessitates time-management. You may want to assign this task to a participant!

A gallery walk is a time-saving process of reporting back from group work. Notes recorded on flipchart for the report-back are displayed on the walls of the room. Participants are invited to walk past the 'exhibits' individually, and read what is recorded. This is followed by a brief plenary discussion. Whenever an activity suggests report-back by gallery-walk, please remind participants to write clearly, using large letters (in lower case for easy reading). The notes should be self-explanatory.

Review and Discussion

This is the most important part of any activity, because it is at this stage that learning and planned transfer of learning happens. Please, always ensure that you have allowed enough time for this part and remember: questions and discussions usually take longer than anticipated!

The **review** affords participants to reflect on the activity just completed; you should consider questions of process (How did the activity go? How did it make you feel?) as much as content (what did you learn about...? What do you think about...?) In most cases the review and discussion section involves many questions; we have listed sample questions — you are encouraged to make your own.

The **discussion** involves an exchange of raw data, as much as views, opinions and experiences. The discussion usually generates a lot of (new) information and you need to allow time so that the new information can be fully understood by participants, and integrated with previous their knowledge.

Summary, assessment and evaluation is the last stage of each activity. This is a useful moment for re-stating what has been dealt with, for naming what has been learnt, for suggesting how the knowledge will be used and applied in the field. All these processes of naming, listing, re-stating, summarising, suggesting etc. are part of assessing whether the purpose of the activity has been achieved. This book does not primarily set out to transmit technical information; hence, evaluation is not based on regurgitation of such information. Facilitators are more concerned with helping participants to assess what they have learnt and how they will use and apply their learning. For suggestions on assessment and evaluation see facilitator's notes.

Important suggestions are usually highlighted as **notes**. These are instructions to you as a facilitator, rather than to the participants.

Grey boxes contain information needed by participants in order to understand the background to an activity. Sometimes the grey boxes have a heading that states sample, or example — in this case we give you an example to illustrate what we mean, and we invite you to formulate your own example.

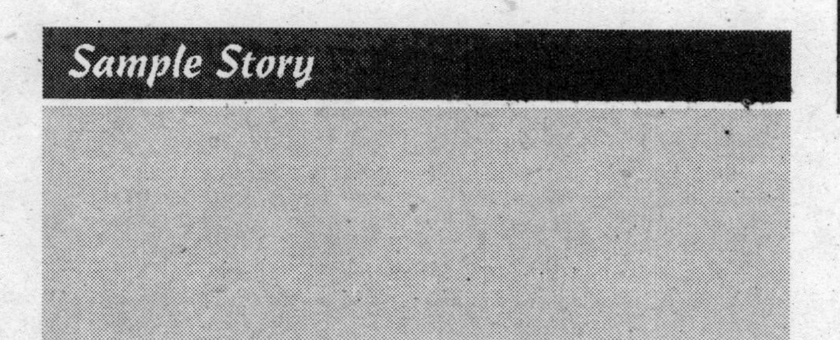

Hints contain further suggestions to you as the facilitator. Hints often include suggestions forfollow-up activities, or references that you may want to consult.

The **resources** section of each activity consists of the materials you need in order to conduct the activity. The symbols in the heading tell you what preparation is needed, eg. Copying and cutting, or enlarging and copying. We suggest you photocopy the resources well ahead of a workshop or course. In some cases materials should be copied on overhead projector transparency. If you do not have such equipment you could copy the materials by hand, on a flipchart.

Hint: you may want to establish a set of cards and materials that you can re-use in subsequent workshops. Store materials in a large envelope and keep them in the back of this book.

Notes for facilitators

Preparing for the workshop

Prepare Materials and Resources

Although activities do not specifically indicate this, we have made the assumption that you have access to flipchart paper or large sheets of newsprint; if this is unavailable you can use old newspapers and write in large letters with a brightly coloured pen. You also need a ready supply of marking pens, preferably in a range of colours. You do not need a special flipchart stand — use the walls of the room for displaying information. Get adequate amounts of sticky stuff — such as blue tack, prestik or masking tape needed for sticking up flipcharts. If your venue has a chalkboard or whiteboard you may want to use those instead of flipchart; in this case check on your supply of chalk or special whiteboard pens!

We have found it useful to always have a box full of stationery and equipment such as additional pens, cardboard or paper cards in different shapes and sizes, scissors, a punch and a stapler. We also give working groups egg-timers so that they can collectively monitor the time.

When preparing for a session you should have access to photocopying facilities; many of the learning activities include resources that need to be photocopied for all participants. Please note that in some cases materials should be enlarged before they are copied; this ensures that all participants can get fully involved in the discussion.

It is very useful to have a small reference library of other training manuals, books and materials that are based on adult education principles, are participatory and experience-based. This should include a dictionary that allows participants to look up words with which they are unfamiliar (see reading list at the end of this book).

Brief Resource People / Subject Specialists

Resource people invited to run a session, or even just give a presentation, need to be carefully briefed: you cannot assume that because they know a lot about a subject, they are familiar with the participatory learning approach you have chosen. A session in which participants feel that their experience is being ignored, and that they are not allowed to build on existing knowledge may do a lot of damage to the general learning climate. Ensure that you instruct resource people clearly: tell them what is expected of them both in terms of content and in terms of the learning/teaching process. Make them aware of what learners already know or have covered in the session, or suggest they begin their session with a brief information-sharing introduction that allows them to gauge available knowledge and experience.

Plan and Design your Learning Sessions

Design your workshop or course in such a way that each session is made up of a mix of different kind of activities. This ensures that participants will not lose interest; it also allows for individual learning styles. Not all participants enjoy group discussions all the time; some may want time for quiet reading and reflection, and all need moments in which they consider how they will apply what they have learnt in their work in the field. Remember that sessions after lunch are often jokingly referred to as 'graveyard sessions', as participants tend to be sleepy and low in energy. Select an active exercise such as a simulation for these sessions — and schedule frequent breaks and energising games for the afternoon. Also remember to keep time, including closing time!

Helping Participants to Learn

You as the facilitator play an important part in establishing and maintaining a learning environment that is conducive to cooperative work and growth. These notes aim to provide some useful hints on how to set up, facilitate and manage a productive learning process. The suggestions given here are general; they apply to all participatory learning but they represent an important shift in the approach to community-based risk reduction.

This section contains the following:

1. an outline of some of the diverse roles a facilitator has to play
2. tools for setting up the learning environment at the beginning of a workshop or course
   - 2.1. suggestions for icebreakers, trust-building and energising games
   - 2.2. information giving, ground rules and learning contracts
3. activities for group formation and maintenance, and participation
4. exercises to improve group communication skills
5. a variety of evaluation methods and processes

1. Roles of the facilitators and participants

This book suggests a departure from the traditional roles of teacher and learners, in that both act as producers of knowledge and developers of skills. Both parties have to agree to the negotiated learning process, and make it happen. Thus, your relationship with the group is one of interdependence: neither you nor they can enact your roles without the participation and assistance of the other. You cannot make the participants learn any more than they can make you teach; you can only facilitate their efforts to learn.

Facilitation is a demanding process; it asks much more of you than to be a teacher or trainer who gives instructions and sees to it that they are followed. As the learning activities in this book imply, you are called upon to play a range of different roles: there are moments when you perform the role of a teacher who offers or explains information; at other times you act as a guide assisting participants through a task; you are often a leader who initiates processes such as group counsellor; at all times you are also evaluator or assessor monitoring and judging a process and making decisions based on your assessment; you act as negotiator and mediator when there are tensions or unhappiness in the group, and a lot of the time your energy is taken up by organising and making practical arrangements.

Similarly, the participants take on different roles as they share in the responsibilities of the learning process. Like you, they act as resource people and teachers, counsellors and leaders, administrators and adjudicators. This is as demanding on them as it is on you because participants have to take responsibility for their own individual learning, as much as the learning and productivity of the whole group.

This process of shared learning does not simply happen by itself; it has to be set up and continuously managed and maintained. All learning groups go through stages and problems and tensions are common. Initially, participants may feel a little reluctant to engage in the learning experience, and they may feel uneasy to speak up in plenary discussions. As the workshop or course progresses, participants improve their sense of confidence and trust, and as people get used to each other, group cohesion develops.

Learning Together

Because participants attend sessions as individuals with a range of experiences, expectations, needs and hopes, the learning group needs nurturing. This is so for a number of reasons:

- Cooperative work depends on a positive sense of group identity, and this should be constructed carefully so that every member is an individual whose needs are being met;
- Participants carry with them, deep-rooted notions of what education and training entails. Their expectations are often based on school experiences, and run contrary to active, co-operative, participatory learning processes in which their own contributions are valued. Initially, they may feel hesitant or even hostile to learning activities that do not run according to their expectations of lecture-style presentations.

2. Setting up a cooperative learning environment and process

The learning environment

The physical environment in which the learning takes place is important. Participants need to feel physically comfortable, so ensure ventilation and light. Participants should be able to see and hear each other; chairs arranged in a circle usually encourage maximum communication. Participants all need to be able to see the flipchart — a mobile stand is useful and a room with plenty of empty walls on which large sheets of flipchart can be displayed is most versatile.

2.1. Ice-breakers, trust building and energising games

Your major function at the beginning of a workshop or course is to create a climate that encourages participants to get to know each other. Good introductions set the stage for a relationship of mutual trust and frankness — basic requirements for productive cooperative learning and confident participation from all. The more comfortable participants are with each other and with you, the more productive and open to learning they are.

Note

Below are some suggestions for different activities; you may want to browse through manuals and books to get further ideas. Most importantly: we have found that participants are amazingly resourceful: ask them for suggestions and try out their ideas — this encourages creative thinking and risk-taking. If the idea works it will fulfil its purpose and be an inspiration to others; if it doesn't: laugh it off and discuss why it didn't and how it could maybe work another time.

The picnic game

This is a name game at the end of which participants will all know each others' names! Sitting in a circle, invite everyone to come to an imaginary picnic to which they should bring something that starts with the same letter as their name. Explain that participants introduce themselves by giving the name by which they want to be called by the group (this is important!), and then an object beginning with the same letter as their name. Start off by giving your name and an object, as an example: 'My name is Astrid and I will bring apples'. The next person in line will say his/her name and the object, and then repeat your name and object. The third person repeats the previous two, and so on. The person who is last has a hard job — but by then everyone will help!

The wind blows on all those who...

This is a great non-threatening energiser in which participants can find out information about each other and get to know each other better. Everyone sits on chairs in a circle. One person stands in the middle. S/he makes a statement that begins with the phrase 'the wind blows on all those who...' (e.g. ... are wearing something blue). All participants fitting that description get up quickly and swop places with each other; the person in the middle tries to get a chair. The person left without a chair introduces the next statement. The game should start with physical observations and then move on to statements about more private matters (e.g. '... has three or more children', or '... works for an aid agency').

Drawing introductions

Give participants sheets of paper and coloured pens. Explain that in order to tell each other more about yourselves, you will draw pictures. Ask participants to write their names at the top of the paper; and underneath draw three things. Depending on what you want to focus on you could ask them to draw pictures of their work, or important moments in their lives, or something that they value very much, or a dream / wish they have. Set a time limit and then ask all participants to present their drawings to each other.

Cultural Greetings

Prepare strips of paper on which you have written different forms of greetings as they are customary amongst different cultures or groups of people. For example: shaking hands, embracing, rubbing noses, bowing etc. Give each participant a strip of paper with a greeting. Now everyone has to move around greeting everyone else in their particular form of greeting. You may want to follow this with a brief discussion on how greetings reflect something about cultural values and traditions and how they all command respect.

Touch Blue

Everyone stands in a part of the room. Explain that you will give instructions that have to be followed as quickly as possible. For example: if you say 'touch blue' (or wood, or hard, or high) all participants have to find and touch something that is blue (or wood, or hard etc.). The last person to follow the instruction has to introduce the next one.

Mirroring in pairs

Participants stand in pairs, opposite each other. One partner makes movements, the other mirrors them. The slower the movements are, the easier it is to reflect them accurately. You might want to begin this activity with giving a simple theme such as 'waking up', or 'getting dressed' or 'preparing a meal'. Swop over so each partner gets a chance to lead and mirror.

Four-up

This can be played with participants remaining around a table or wherever they are sitting. The aim of the game is that at any stage four players stand up for anything up to 15 seconds. As they sit down someone else gets up to complete the four.

Knots

Participants stand in a circle (no more than 8 per circle). They all stretch their hands into a circle and each person grabs and holds on to two hands belonging to different people across the circle. Ask everyone to stand back without letting go of the hands. The aim of the game is to get back into the circle without letting go of any hands. The process of disentangling the 'knot' may involve stepping over or under etc — but it always requires cooperation.

2.2. Information Giving, Ground Rules and Learning Contracts

Adult learners are expected to take charge of their learning, but they need to feel secure that they know what is expected of them. This requires firstly, that they have the necessary information and secondly, that the process mechanisms allow them to do that. The act of establishing contracts and monitoring groups provides a useful model for participants' own work in the field. Ensure that they will be able to transfer the insights and skills from this process to their work in risk reduction at community level; facilitate discussions about the process and highlight the decision-making steps you went through.

Session outlines and programme negotiations

Item number one of any programme should be a collective review of the proposed outline of the course/session: even if participants have been sent/ handed the outline in advance discuss it together and give a clear message that this is a proposed outline that is open to negotiations. Take participants through the agenda, explain how you envisage the course/session to run; encourage participants to interrogate the agenda by asking questions and challenging suggestions. Affect proposed and discussed changes immediately.

At the beginning of each day, review the agenda: reassess whether there is a common understanding of what will happen during the day. Be flexible enough to introduce changes should the need arise and negotiate suggested deviations from the agenda.

Towards the end of the course/session include a time slot for future planning. Review evaluations and learning assessments conducted, and ask for proposed items for the next course/session

Participants' Hope and Fears

In most cases we do not know what exactly to expect when we attend a training or learning workshop. Our expectations may be totally different from what will actually happen. We have found it most useful to begin a session by asking participants to list their hopes and fears, rather than their expectations. Providing cards on which those hopes and fears can be written allows us to collectively sort them into groups of issues/topics and eliminate duplicates. As a group, we then address those issues — providing information, addressing worries and concerns, and encouraging discussions that lead to greater clarity and understanding. This process is an essential part of trust-building in the group. An affirming way to conclude the process is by requesting all participants to suggest ways of ensuring that the hopes are fulfilled!

Setting Ground Rules

Once you have established the hopes and fears, and collected suggestions for ways in which you plan to realise those hopes you can move to setting ground rules. These are the 'laws' that guide the process of collective learning. Examples include rules about smoking, punctuality, clear use of disaster related terms and concepts, non-sexist language etc. An easy way of setting them up is by asking participants to list things that each one can do that would ensure that the learning process will be productive and enjoyable. Suggestions should address issues of personal conduct and inter-personal relations. Write up the ground rules and display them clearly visible in a part of the room. Include suggestions of how you will deal with 'offences'.

Learning Contracts

Once rules have been identified, and the purpose and process of the disaster reduction learning course has been clearly established, it is helpful to enter into a binding contract with the participants. A 'learning contract' should be negotiated and drawn up in collaboration with participants; it sets out clearly what is expected of both learners and facilitators in terms of personal conduct and contribution to the course, and at best it spells out how the process will be monitored and assessed.

Process Monitoring Group

There are different approaches to the functions and workings of a process monitoring group. One might be that participants work in small groups and take turns to draft and administer daily evaluations. The findings are reported daily to the large group. Another way has pairs of participants taking turns in recording proceedings; in this way participants contribute to producing a collective journal on 'learning about risk reduction'; records are checked in morning plenary sessions and all participants receive copies of the journal which they can use for reporting back to their organisations after the training.

Discuss and negotiate the functions of the process group; these may range from time-keeping to record-keeping, from tasks related to the distribution of information, to working as 'mood barometers' that take care of emotional and physical well-being of all participants. If you as the facilitator have the main responsibility for the learning programme the monitoring group is mainly in charge of participants' well being.

3. Group Formation, Maintenance and Participation

Group Formation

There are many ways of dividing participants into small working groups. The simplest way is 'counting off' to the number of groups you need to create: the first participant is 'one', subsequent participants are 'two', 'three' and 'four', and then you begin again with 'one', etc. A more creative way is to cut a number of pictures that illustrate vulnerability or capacity into *puzzle pieces*; give each participant one piece and ask them to find those participants who hold the other pieces of their picture. When they have found each other they could have a brief discussion around the vulnerability / capacity illustrated.

Number groups is a more active way that combines an energising game with group formation. Ask participants to stand anywhere is the room; call out a number, for example 'three' and ask participants to get into groups of three. Then call out another number and ask them to get into groups of that number, and another, and finally the number of the group size that you are aiming at.

Resourceful Creations

This activity asks participants to make a beautiful boat that floats. Participants work in small groups of 4-5; each group is given a number of materials such as a sheet of paper, some crayons, five paper clips, three toothpicks; set a time limit and ask groups to use only the materials given and create a boat. Point out that the boats will be judged according to two criteria: their ability to float, and their aesthetic appeal. At the end of the time ask groups to float their boats in a bowl of water and collectively judge the boats. Review and discuss the process: how did they go about planning their creation? Did they assess the materials given? (e.g. the fact that wax crayons are water-resistant). How did they work together as a group? How were decisions made and implemented? What resources did they draw on in order to create the boat? What assisted/hindered cooperation? Draw general conclusions about creative collective problem-solving and planning. How does this activity relate to disaster mitigation?

Cooperative Drawings

Participants work in pairs; they have a piece of paper and one pencil between them. Instruct them to both hold on to the pen, and without communicating with each other in any way, draw a building. This will usually generate a lot of laughter and noises of frustration. Set a time limit; display the drawings on the wall and initiate a review of the process, and discussion around cooperative work. What does it take for partners to work collectively? What happens to leadership? How are decisions made? Pick out drawings that illustrate good or poor cooperative work. Ask the couples to describe how they went about drawing their building (ensure both partners get a chance to present their view of the experience!). Draw general conclusions about cooperative work and leadership roles in disaster mitigation. How are these important for maintaining a productive working group?

Talking Beans

This is usually a very effective and enjoyable process that demonstrates convincingly what it takes for us to become more considerate listeners and group participants. Ask participants to get into small discussion groups and give each group member about three to five beans (or other such tokens). Instruct them to have a discussion (preferably around a contentious issue!). Each speech-act 'costs' one bean — irrespective of what kind of speech-act it is: interjection, question, statement of agreement or substantive information/opinion giving. Once all 3-5 beans have been spent the person can no longer contribute to the discussion. When all group members have 'spent' their beans review the process: what happened? On what kind of contribution to the discussion did participants spend their beans? Did they feel any were wasted?

Discuss what insights about speaking/listening participants gained; what speaking and listening strategies did they develop (e.g. keeping one bean to be the last speaker)?

Sabotage

Conversations are often hijacked or sabotaged by the acts of one person. This may not be deliberate — but the effect is the same. This activity discovers how it happens, and leads to suggestions of how to deal with the problem. Ask participants to work in threes; they are to take turns in acting out the parts of talker, listener and saboteur. Instruct them to tell each other stories of something interesting and exciting that happened. Discuss which of the roles was easiest / most difficult to play. Why? Discover different ways of sabotaging a conversation, and various means of dealing with it. Write up suggestions / strategies in which participants have dealt with sabotage.

4. Communication Skills

Listening Skills

Point out that listening is the most important communication skill of all. Listening means creating the space for someone to speak and simply be listened to, without any comment, interruption or judgement. There are many listening exercises — this one is a *group story-telling* process that is fun. Sit in a circle; suggest a topic related to an emergency situation (e.g. 'the day the rains came' or 'when Thandi's family moved away') and begin to tell a story. Do not contribute more than about three sentences, then ask the person next to you to continue the same story, and so on, around the circle. Participants should aim to develop incidents and characters and make the story interesting. Discuss what happened. Focus on what helped or hindered listening.

So you know it well?

This brief exercise demonstrates how we often think we know a lot about things because they are familiar to us. Ask for two volunteers; take their watches (or other familiar objects, such as shoes) from them. Ask them to sit with their backs to the rest of the participants and instruct them to take turns in describing their watches so that they are easily recognisable. Unpack what happened, and why/how we often do not look closely at something that is familiar. Ask participants to relate this demonstration to problems we have to confront when conducting risk assessments: what does this tell us about different perceptions of risk?

Fact opinion rumour

This activity serves to demonstrate the difficulty we face when distinguishing facts from opinions and rumours. Read out a simple text — such as a brief newspaper article; pause after each statement. Ask participants to consider whether the statement was a fact, opinion, or rumour. If they decide it was a fact, they are to raise their hand in the air; if it was an opinion they should put their hand on their head; if it was a rumour they should fold their arms. This activity demonstrates how we should not assume that everyone understands and thinks the same way as we do. Facilitate a discussion around the different ways in which we perceive things. Therefore, we cannot simply accept whatever we hear but we have to assess and judge it.

Following instructions

Communication is not as straight as it seems — this becomes very clear when we are asked to follow simple instructions. (a) Ask for four volunteers and instruct them to stand in front of the class. Give each a sheet of paper and ask them to close their eyes. Now request them to follow the instructions you will give. Ask them to fold the paper in a number of ways, end off by askingthem to tear off a corner. When the papers are unfolded it becomes clear that interpretations of the instructions were different. What can we do about ambiguity? (b) Stand in front of the class and take off your jacket or jersey. Ask participants to give you instructions that will end in you wearing the jersey. Follow instructions very literally; this will usually lead to a muddle and not a jersey or jacket worn properly!

5. Ongoing Evaluations

In any learning process there are many different things that need to be evaluated on an ongoing basis, for example the learning: whether participants are learning what they want and need to learn, whether they are feeling confident and comfortable with the process, whether they are able to apply what they learn in their work in the field. It is also important to know why something is successful, or why and how it fails to achieve what it sets out to do. We have found it useful to conduct small daily evaluations to assess learning achievements and participants' well-being, and more detailed assessments at the end of each training block and in between sessions.

Such evaluations took many different forms; these are just a few examples. Common to all is the need to know more than whether participants liked a session. Ensure that you ask open questions that challenge respondents to reflect on *how* they liked something, and name exactly *what* they learnt. In this way you gather important information for planning further activities and sessions.

Remember: the general climate and mood amongst all participants (including facilitators) is the best indicator of a successful workshop — the details of how useful the learning was will only be revealed back at work, in the field.

Quick Assessment Procedures

Assessments can be quick procedures conducted orally — and the activities in this book contain a number of such examples. Alternatively, you may want to ask participants to respond individually and anonymously, in writing. Ensure that you *identify clearly, what you are assessing*: for example, the experience created by an activity, or the facilitator's ability to manage the process, or the clarity of presentation, or the participants' understanding, or the choice of focus and topic, etc. In order to get as much quality information as possible, ask open questions, such as 'how does the information on drought improve your understanding of environmental degradation?', or 'what insights have you gained as a result of the simulation?'.

One way of getting very focused, precise responses is by formulating questions as open unfinished statements, for example 'I now know that perceptions of risk are ....', or 'My understanding of vulnerability has improved because I can now see that ....', or 'I now realise I need to further develop my ability to ....'. In written evaluations, remember to leave some space for comments.

Detailed Impact Assessments

Detailed written individual assessments are useful tools that allow participants to gauge improvements in their ability to initiate and affect risk reduction interventions. Participants respond in writing to questions and challenges that ask them to reflect on what aspects of the learning programme they had found particularly useful, and why, and how they had applied what they learnt in the field. If they had not been able to use their new learning — why not?

Work Experience Reports

These are reports prepared by participants in advance of a course/session. A while before the next session inform participants that they will be

asked to present work experience reports. Outline specific questions that you would like them to respond to: the more specific the instruction are the easier it is for participants to respond and prepare something that will be focused and relevant. Include an outline of the process which the presentations will take. At the end of the presentations review some of the common difficulties and problems: investigate how participants dealt with the problems, highlight similarities and differences and ensure that there is a clear link between the work reports and the disaster mitigation workshops. In this way everyone involved can see how useful the course has been, and what gaps remain to be filled.

Remember: the main purpose of an evaluation is to help you plan for future work.

Section 1:

Key Concepts and Risk Factors

This section contains ten learning activities that are mostly conceptual in nature, but practical in application. They aim at developing participants' understanding of concepts and terminology with regard to risk, both as it shapes the daily lives of people and communities, and as it impacts on the work of non-governmental and humanitarian organisations. The activities in this section 'set the stage' for further learning activities, as they stimulate critical thinking around issues of risk reduction as an essential component of development.

What terms do we use in disaster management?

Purpose

This activity aims to build a common understanding of 'disaster management' terminology.

Note

Participants come to the training with different assumptions and understanding of what disaster management terms mean.

Procedure

In this activity participants begin by giving their own definitions of terms such as 'hazard', 'emergency preparedness', 'mitigation' etc. This is followed by discussions until the group reaches consensus on the meaning and definition of terms.

Time

- 1 hour

Materials

- glossary of disaster management terms as a reference and possible post-session hand-out (see resources)

Process

Introduction

1. Introduce the activity by outlining the purpose and procedure.

Participant Action

1. Ask participants to reflect briefly on some of the key terms they associate with 'disaster management'.

2. Ask participants to name those key words. Write each word on a separate sheet of flipchart paper.

3. Stick flipcharts on the walls.

4. Invite participants to come forward and write brief descriptions explaining those key words on the flipcharts.

Encourage many different suggestions from as many of the participants as possible and collect a broad spectrum of words. Do not 'censure' any suggestions at this stage.

Review and Discussion

1. In plenary, facilitate a discussion around each of the descriptive words and phrases written on the flipchart paper:
   - Ask for clarification and examples.
   - Ensure that meanings of terms are clearly differentiated.
   - Mark those words that are crucial to the definition.
   - Agree on a broad definition of each term.

2. Ask participants to work in pairs. Ask each pair to summarise the discussion by writing one of the terms and its agreed-upon definition on flipchart paper.

3. Review the 'glossary on flipchart' in a gallery walk.

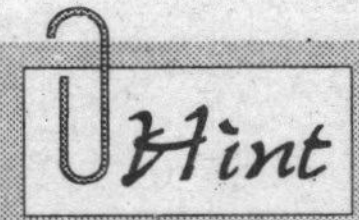

You could suggest that this list of terms and definitions on flipcharts should be displayed as a reference throughout the course. This will facilitate continued common understanding of terminology.

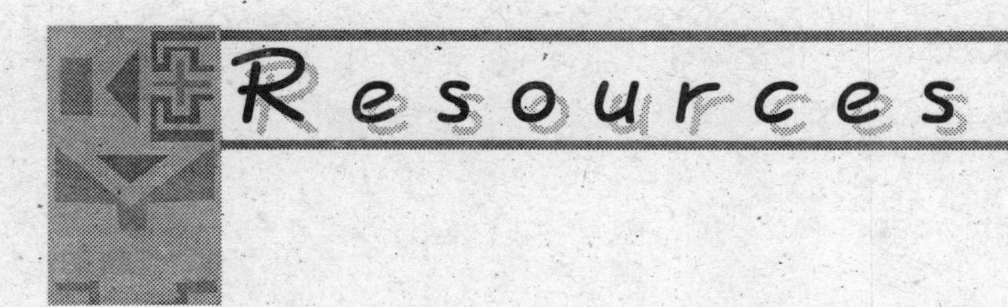

GLOSSARY OF DISASTER MANAGEMENT TERMS[1]

Disaster management

A collective term encompassing all aspects of planning for and responding to disasters, including both pre- and post-disaster activities. It may refer to the management of both the risks and consequences of disasters.

Disaster

A serious disruption of the functioning of a society, causing widespread human, material, or environmental losses which exceed the ability of the affected society to cope using only its own resources. Disaster is sometimes also used to describe a catastrophic situation in which the normal patterns of life (or eco-systems) have been disrupted and extraordinary, emergency interventions are required to save and preserve human lives and / or the environment. Disasters are frequently categorised according to their perceived causes and speed of impact.

Human-made disaster

Disaster or emergency situation of which the principal, direct causes are identifiable human actions, deliberate or otherwise. Apart from "technological disasters" this mainly involves situations in which civilian populations suffer casualties, losses of property, basic services and means of livelihood as a result of war, civil strife, other conflict or policy implementation. In many cases, people are forced to leave their homes, giving rise to congregations of refugees or externally or internally displaced persons.

Slow-onset disasters

(also called Creeping Disasters or Slow-onset Emergencies)

Situations in which the ability of people to sustain their livelihood slowly declines to a point where survival is ultimately jeopardised. Such situations are typically brought on or precipitated by ecological, social, economic or political conditions.

Sudden-onset natural disasters

Sudden calamities caused by natural phenomena such as earthquakes, floods, tropical storms, volcanic eruptions. They strike with little or no warning and have an immediate adverse effect on human populations, activities, and economic systems.

Technological disasters

Situations in which large numbers of people, property, infrastructure, or economic activity are directly and adversely affected by major industrial accidents, severe pollution incidents, nuclear accidents, air crashes (in populated areas), major fires, or explosions.

Hazard

(also called Hazardous Phenomenon or Event)

A rare or extreme natural or human made event that threatens to adversely affect human life, property or activity to the extent of causing disaster. A hazard is a natural or human-made phenomenon which may cause physical damage , economic losses, or threaten human life and well-being if it occurs in an area of human settlement, agricultural. or industrial activity. Note, however, that in engineering, the term is used in a more specific, mathematical sense to mean the probability of the occurrence, within a specified period of time and a given area, of a particular, potentially damaging phenomenon of a given severity or intensity.

Human-made hazard

A condition which may have disastrous consequences for a society. It derives from technological processes, human interactions with the environments, or relationships within and between communities.

Natural hazard

Natural phenomena which occur in proximity and pose a threat to people, structures or economic assets and may cause disaster. They are caused by biological, geological, seismic, hydrologic, or meteorological conditions or processes in the natural environment.

Hazard assessment

(also called Hazard Analysis or Evaluation)

The process of estimating, for defined areas, the probabilities of the occurrence of potentially-damaging phenomena of given magnitudes within a specified period of time.

Hazard assessment involves analysis of formal and informal historical records, and skilled interpretation of existing topographical, geological, geomorphologic, hydrological and land-use maps,as well as analysis of social and economic and political conditions.

Hazard mapping

The process of establishing geographically where and to what extent particular phenomena are likely to pose a threat to people, property, infrastructure, and economic activities. Hazard mapping represents the results of hazard assessment on a map, showing the frequency / probability of occurrences of various magnitudes or durations.

Risk

Risk is defined as the expected losses (lives lost, persons injured, damage to property, and disruption of economic activity or livelihood) caused by a particular phenomenon. Risk is a function of the probability of particular occurrences and the losses each would cause. Other analysts use the term to mean the probability of a disaster occurring and resulting in a particular level of loss. A societal element is said to be 'at-risk' or 'vulnerable', when it is exposed to known hazards and is likely to be adversely affected by the impact of those hazards if and when they occur. The communities, structures, services, or activities concerned are described as "elements at risk".

Risk assessment

(sometimes called Risk Analysis or evaluation)

Risk analysis is a process of determining the nature and scale of losses and damage due to disaster which can be anticipated in particular areas during a specified time period. Evaluation of risk is the social and political judgement of various risks by the individuals and communities that face them. This involves trading off perceived risks against potential benefits and also includes balancing scientific judgements against other factors and beliefs.

Risk mapping

The presentation of the results of risk assessment on a map, showing the levels of expected losses which can be anticipated in specific areas, during a particular time period, as a result of particular disaster hazards.

Vulnerability

The extent to which an individual, community, sub-group, structure, service, or geographic area is likely to be damaged or disrupted by the impact of a particular disaster hazard.

Vulnerability analysis

The process of estimating the vulnerability to potential disaster hazards of specified elements at risk. For engineering purposes, vulnerability analysis involves the analysis of theoretical and empirical data concerning the effects of particular phenomena on particular types of structures. For more general socio-economic purposes, it involves consideration of all significant elements in society, including physical, social and economic considerations (both short- and long-term), and the extent to which essential services and traditional and local coping mechanisms are able to continue functioning.

Emergency

An extraordinary situation where there are serious and immediate threats to human life as a result of disaster, imminent threat of a disaster, cumulative process of neglect, civil conflict, environmental degradation and socio-economic conditions. An emergency can encompass a situation in which there is a clear and marked deterioration in the coping abilities of a group or community.

Preparedness

Measures to ensure the readiness and ability of a society to forecast and take precautionary measures in advance of an imminent threat, and to respond to and cope with the effects of a disaster by organising and facilitating timely and effective rescue, relief and appropriate post-disaster assistance.

Food security

Access by all people at all times to enough food for an active, healthy life. Its essential elements are availability of food and ability to acquire it. The FAO definition of food security includes the following requirements: adequate supply, stable supply, and access to the supply (including adequate consumption, adequate income in relation to food prices and access to employment).

Food insecurity, in turn is the lack of access to enough food. There are two kinds of food insecurity: chronic food insecurity which results in a continuously inadequate diet, and acute food insecurity which is a temporary decline in a household's access to enough food.

Post disaster assessment

(sometimes called Damage and Needs Assessment)

The process of determining the impact of a disaster or events on a society, the needs for immediate, emergency measures to save and sustain the lives of survivors, and the possibilities for expediting recovery and development.

Assessment is an interdisciplinary process undertaken in phases and involving on-the-spot surveys and the collation, evaluation and interpretation of information from various sources concerning both direct and indirect losses, short-and long-term effects. It involves determining not only what has happened and what assistance might be needed, but also defining objectives and how relevant assistance can actually be provided to the victims. It requires attention to both short-term needs and long-term implications.

Disaster mitigation

Mitigation refers to measures which can be taken to minimise the destructive and disruptive effects of hazards and thus lessen the magnitude of a disaster. Mitigation measures can be of different kinds, ranging from physical measures such as flood defences or safe building design, to legislation, training and public awareness. Mitigation is an activity which can take place at any time: before a disaster occurs, during an emergency, or after disaster, during recovery or reconstruction.

Disaster preparedness

Measures that ensure the readiness and ability of a society to (a) forecast and take precautionary measures in advance of an imminent threat (in cases where advance warnings are possible), and (b) respond to and cope with the effects of a disaster by organising and delivering timely and effective rescue, relief and other appropriate post-disaster assistance. Preparedness involves the development and regular testing of warning systems (linked to forecasting systems) and plans for evacuation or other measures to be taken during a disaster alert period to minimise potential loss of life and physical damage; the education and training of officials and the population at risk; the establishment of policies, standards, organisational arrangements and operational plans to be applied following a disaster impact; the securing of resources (possibly including the stockpiling of supplies and the earmarking of funds); and the training of intervention teams. It must be supported by enabling legislation.

1. These definitions come from a number of sources including United Nations Development Programme (UNDP), Food and Agriculture Organisation (FAO), Andrew Maskrey (Disaster Mitigation - A Community Based Approach), with some modifications and additions.

What are the components of disaster management?

Purpose

This activity aims to clarify the differences between the key elements of disaster management, and their interconnectedness.

Note
This activity draws out participants' conceptions of what disaster management is.

Procedure

Through a process of creating 'spider-diagrams' participants distinguish between key concepts and activities in disaster management/reduction.

Time

- 1 hour

Materials

- blank strips of paper or card

Introduction

1. Introduce the activity by outlining the purpose and procedure.

2. Ask participants whether they are familiar with the meaning of 'topic webs' or 'spider-diagrams'.
   - Explain that this is a useful process for drawing out and organising information such as the various elements of disaster management.
   - If any participants are familiar with 'spider-diagrams' ask them to explain and briefly illustrate what a simple 'spider-diagram' might look like.

Participant Action

1. Give each participant three or more blank strips of paper.

2. Give the following instruction:

> *Write the key words you consider to be central to 'disaster management' on the strips of paper; for example 'mitigation'. (one concept per paper)*
>
> *(allow approximately 5-10 minutes for this)*

3. Write the words 'disaster management' on the centre of the board or newsprint. Draw a box around the words.

4. Ask participants to come forward and take turns in sticking their strips of paper next to / around the words 'disaster management'.

Review and Discussion

1. In plenary, review the arrangement of words. Ask questions such as the following:
   - **?** Should the words be in any specific order?
   - **?** Which words belong together and how should they be arranged to show this relationship?
   - **?** Which words refer to **concepts** in disaster management, and which describe specific **actions** /activities (eg 'planting woodlots' or 'distributing seeds')?

2. Move words describing **activities** into a separate list entitled 'disaster management activities'.

3. Remove duplicate words. Move any words that do not fit where they have been stuck.

4. Ask participants to establish links between the words by drawing lines denoting relationships.

5. Ask participants to identify the area of disaster management in which they work predominantly.

6. Ask participants to copy the 'mind-map' as this will provide a useful reference and record of their learning.

This is an example of one type of web / diagram, showing groupings and linkages.

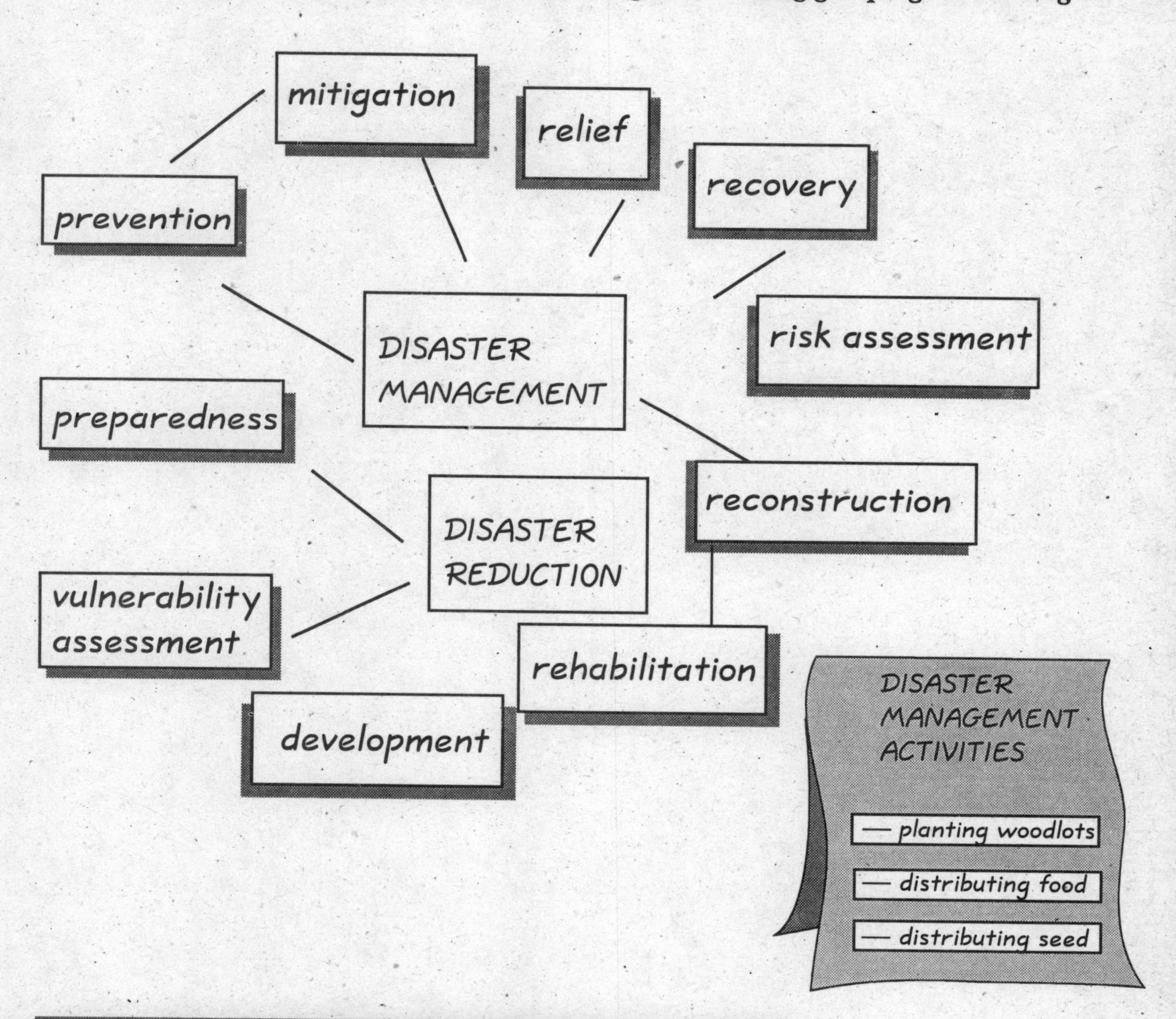

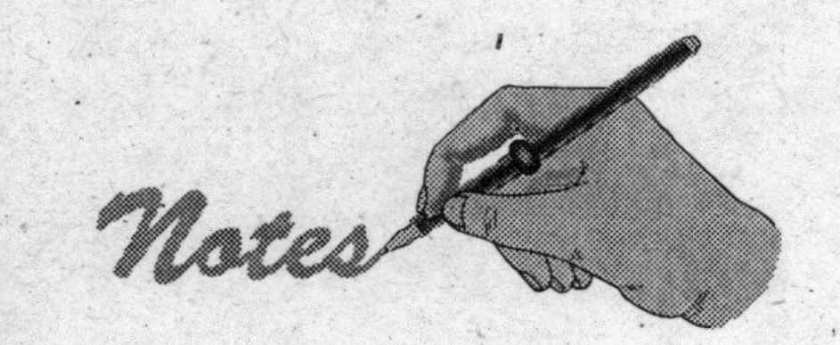
Notes

What do we mean by hazards, risks, vulnerabilities and capacities?

Purpose

This activity asks participants to apply their understanding of key terms and concepts in disaster management to specific hazards.

Note

People deal with risks in their daily lives and they use 'disaster language' to describe mishaps and accidents. However, careless use of terms leads to miscommunication.

Procedure

This activity begins with a simple story that illustrates disaster terminology. After a discussion about how vulnerability increases risk, participants are asked to develop their own stories that illustrate specific hazards in people's daily lives.

Time

◆ 1½ - 2 hours

Process

Introduction

1. Introduce the activity by outlining the purpose and procedure.

2. Tell a brief story which illustrates the effects of a common hazard on a number of people in a daily life situation. In your story use disaster terms such as hazard, risk, vulnerability and capacity.

Sample Story

MaMhlongo is a pensioner who lives quite close to the tar road that leads to town. She is happy that the walk from her house to the road is not far. But she is always worried about that little stretch that she has to walk along the road to the busstop because the minibus taxis zoom past at great speed. 'Those buses are really a hazard,' she was telling her daughter: 'Only last week another child was killed when a minibus swerved to avoid a goat.' And the weekend before she had to attend the funeral of her neighbour who had been knocked dead by a minibus. But this is the shortest way to get to the busstop and these days she walks so slowly. So she rather takes the risk along the road, than the risk of missing the bus by going the long way through the bush. Those paths always pain her legs. Of course, if she were young and healthy like her daughter, she would not be so vulnerable - she could easily just jump out of the way if a minibus came....

3. Ask participants to identify the *hazards, risks, vulnerabilities* and *capacities* referred to in the story. You may want to write the terms and examples given on newsprint.

4. Ask questions to further define hazard, vulnerability, risk and capacity referred to in the story. You may ask questions such as the following:

   ? Does this hazard pose an equal risk to all the people exposed to it? If no, why not?

   ? What makes some people more / less vulnerable?

   ? In this story, how do you think this risk is perceived by the different people affected? Are there differences? Why?

5. Write up and explain the following equation, referring to the story as an example:

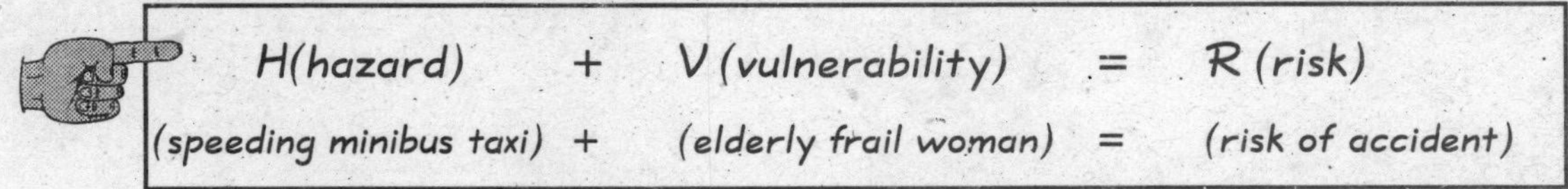

H(hazard) + V (vulnerability) = R (risk)
(speeding minibus taxi) + (elderly frail woman) = (risk of accident)

Participant Action

1. Ask participants to get into small groups of 3-4 people.

2. Introduce the next activity: participants are asked to apply their knowledge of terminology to specific case scenarios.

3. Instruct each group to do the following:

Develop a simple story that illustrates a common risk in a familiar community. Show how vulnerability makes a difference in terms of real risk and perceived risk. How do people cope with it? What capacities do they draw on? Allow 20 minutes for this activity.

This activity was developed on the basis of a story told by one of the participants in the SADMTP course. You may want to tell this story as an example of what the given task involves.

In order to illustrate 'hazard', 'risk' and 'capacity' he told the following tale:

"In a forest near the little village of Donga lives a dangerous snake. Whenever the people of the village would pass through the forest in order to get to the river the snake would drop down on them, and bite them, and they would die. So the people of the village always taught their children to carry a big stone on their head, so that the snake might not bite them."

Review and Discussion

1. Call participants together and ask each group to share its tale.
2. Review the stories:
   - — Identify the given hazards, vulnerabilities, risks and capacities.
   - — Ask for responses: what did participants like / dislike?
   - — Ask for suggestions on how the stories could be improved, for example; made more precise, culturally acurate, relevant, etc.
3. Initiate a discussion around risk: why do perceptions of risk differ from person to person, community to community, society to society?
4. Ask a participant to sum up what s/he learnt from the activity.

How do we check our understanding of terms?

Purpose

This activity aims to build confidence and understanding in the use of key disaster management terms.

Note

This process allows participants to check and improve their understanding of disaster management terminology and concepts.

Procedure

Participants work with definitions of various terms, written on cards. The process of trying to match those definitions with key terms involves differentiated thinking and leads to lively debate.

Time

- 1 hour

Materials

- a set of definitions of disaster terminology, taken from different sources and written on individual cards (see resources)
- copies of those definitions on OHP slides
- key terms written on flipchart paper or cards displayed on the wall (see point 4)

Process

Introduction

1. Outline the purpose and procedure of the activity.

   Point out that disaster management is a relatively young field and definitions of key terms given in the literature often differ. This causes confusion.

2. Distribute cards with written definitions amongst the participants.

3. Suggest that participants work in pairs. Give the following instruction:

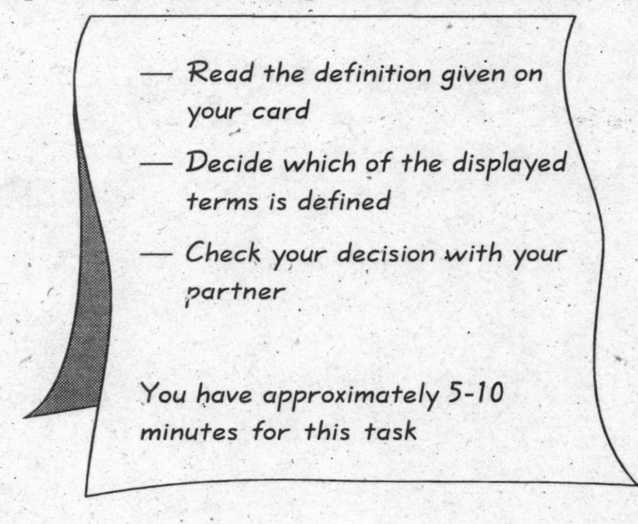

— Read the definition given on your card
— Decide which of the displayed terms is defined
— Check your decision with your partner
You have approximately 5-10 minutes for this task

4. Meanwhile, arrange and display the key terms on the wall, in the following way:

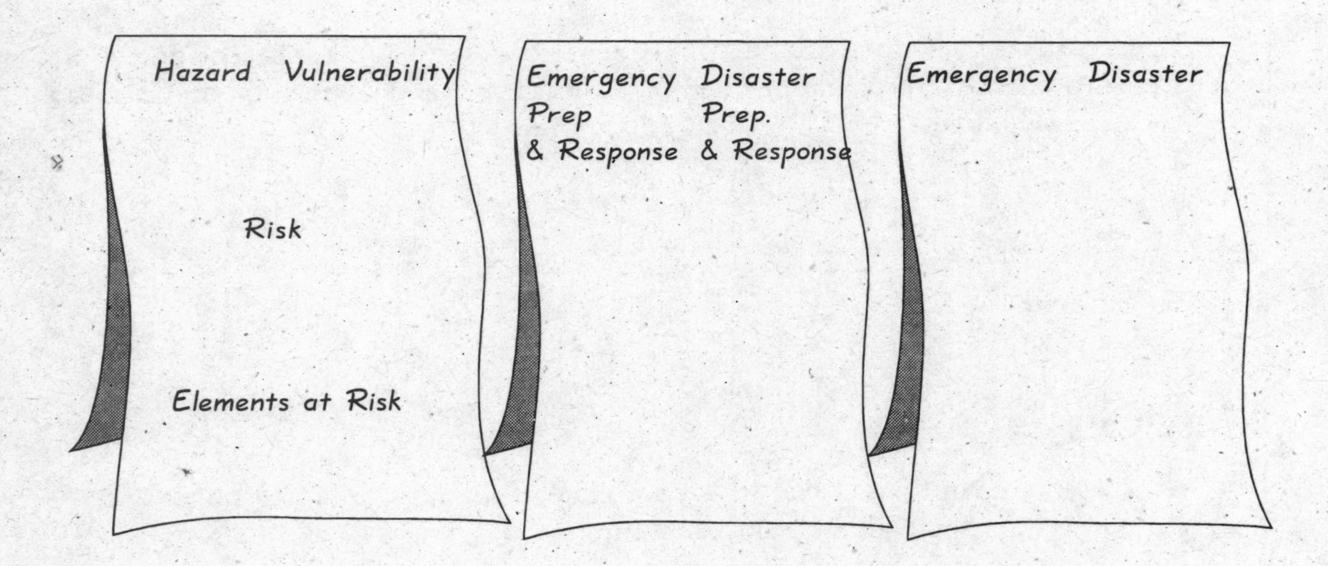

Participant Action

1. Participants spend 5-10 minutes on their task.

Review and Discussion

1. Point to the first term: hazard, and ask which participant thinks s/he has the card with the definition of 'hazard'.

2. Ask the participant who indicates s/he has the appropriate definition to read out her/his card.

   Check whether other participants agree with this being a definition of 'hazard'.

3. In case of disagreement: display the given defintion on OHP slide, in order to allow all participants to read what is written on the card. This facilitates participation in discussion. Encourage debate and contestation.

4. Manage the discussion. If necessary, explain the key difference between one term/ concept and another.

5. When participants have reached consensus stick the card underneath the term displayed.

6. Proceed in this way with all other terms / definitions.

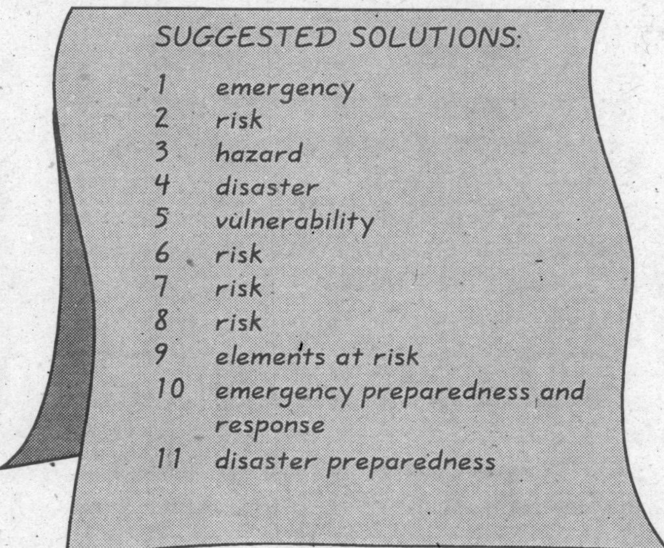

SUGGESTED SOLUTIONS:
1 emergency
2 risk
3 hazard
4 disaster
5 vulnerability
6 risk
7 risk
8 risk
9 elements at risk
10 emergency preparedness and response
11 disaster preparedness

7. Sum up the activity by reminding participants that the definitions are guidelines rather than cut and dry 'rules' as the understanding and usage of terms and concepts are flexible and changing as the field of disaster management develops.

You could ask participants to write examples of key emergencies, hazards, risks, etc. on cards and ask each other to classify them in the same way.

Notes

Resources

DEFINITIONS OF DISASTER TERMINOLOGY

1

A situation where there are serious and immediate threats to human life because of

-a disaster

-threat of a disaster

-process of neglect

-civil conflict

-environmental degradation

2

This term refers to the expected impact of a given element at risk over a future specified time period.

3

An event with the potential to cause injury or death, or to damage property or environments on which people depend.

4

A serious disruption of the functioning of a community causing widespread human, material or environmental losses which exceed the ability of the affected community to cope using its own resources.

5

A condition or set of conditions which reduces people's ability to prepare for, withstand or respond to a hazard.

6

The degree of negative change anticipated when hazard(s) occur under conditions of vulnerability.

7

The extent to which a community, structure or geographic area is likely to be damaged or disrupted by the impact of a particular hazard, on account of their nature, construction, and proximity to a hazardous area.

8

HAZARD x VULNERABILITY = ?

9

PEOPLE
PROPERTY
ECONOMY
SOCIAL SERVICES
ENVIRONMENT

10

An ability to predict, respond to and cope with the effects of an emergency, to reduce its effects and ensure timely and appropriate response

11

The ability to predict, respond to and cope with the effects of a disaster.

How does the 'disaster crunch model' work?[1]

Purpose

This activity aims to clarify how the link between hazards and vulnerabilities increases the risk of disaster.

Note

Participants consider how vulnerabilites lead to increased risk, and how key interventions can avert disasters.

Procedure

This activity begins with a reflection on the impact of hazards on participants. It then moves to a more theoretical discussion based on the 'disaster crunch model'. Finally, participants test the proposed model by applying the components to specific case scenarios in southern Africa.

Time

- 90 minutes

Materials

- copies of 'The Disaster Crunch Model' for each participant, and on OHP slide (see resources)
- pens and paper

Process

Introduction

1. Outline the purpose and procedure of the activity.
2. Remind participants that the aim of this course is to switch the focus from disaster management as emergency preparedness and response, to disaster reduction as the cornerstone of development. Explain that 'hazards' and 'disasters' are not synonymous: not every hazard leads inevitably to a disaster. Suggest that development workers need to have a clear understanding of how the impact of a hazard can lead to a disaster. Some programmes can also increase or decrease disaster risk.
3. Refer to the 1991/2 drought (or a similar recent emergency in the region). Ask participants to describe the impact of the drought on them personally.
4. Explore reasons why the event had a serious effect on them, or why it did not?

   Discuss the environmental, social, economic impact; who was most affected, and why?
5. Introduce the 'disaster crunch model' (also known as the 'pressure and release model') as an example of how particular conditions and causes give rise to 'the progression of vulnerability', and hence to potential disasters (see resources).
6. Take an example of a hazard common to Southern Africa, such as a drought, flood, storm, dysentery etc. Analyse it, using the 'crunch model' as a guideline. Explain the forces described in the model.
7. Encourage participants to ask questions of clarification.
8. Check for understanding of concepts by asking questions such as the following:
   - **?** What is meant by 'limited access to power structures'?
   - **?** What are examples of 'environmental degradation'?
   - **?** What places 'livelihoods at risk'?

Participant Action

1. Ask participants to get into groups. Explain that they should now apply their knowledge to examples from their experience.

2. Give the following instruction:

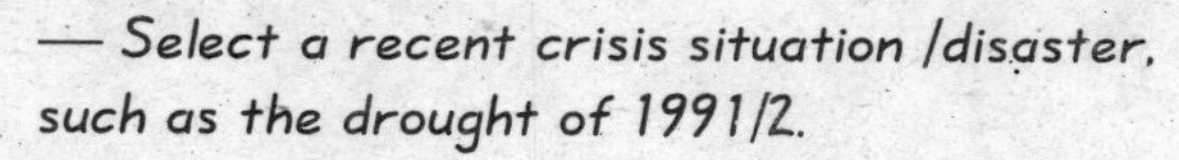

— Select a recent crisis situation /disaster, such as the drought of 1991/2.

— Using the 'disaster crunch model' as a guideline refer to the headings listed under 'progression of vulnerability'.

— Analyse the crisis situation/disaster in terms of causes, pressures and conditions, and their relationship to specific trigger events.

— Prepare a brief presentation of your findings: use flipchart paper placed horizontally to reproduce your example like the 'disaster crunch model'.

You have 30 minutes to complete the task

3. Monitor the progress and assist groups, where necessary.

Review and Discussion

1. Facilitate presentations in plenary.

   After each presentation allow a few minutes for questions of clarification and/or challenges to the presenters' analysis.

2. Initiate a plenary discussion on questions such as the following:

   ? What have we learnt from this process? What were the surprises?

   ? What were similarities and differences for different hazards?

   ? Were there common vulnerabilities across different hazards?

   ? How could we use an analysis of vulnerability (causes, pressures and conditions) as a warning system?

   ? Given an analysis of the present conditions in our country / area / region, what hazards are particularly threatening?

14. Ask participants to critically review the 'crunch model' as a diagram: how does it help them to better understand disasters?

15. Summarise the findings.

This activity should be followed by a more detailed vulnerability assessment procedure.

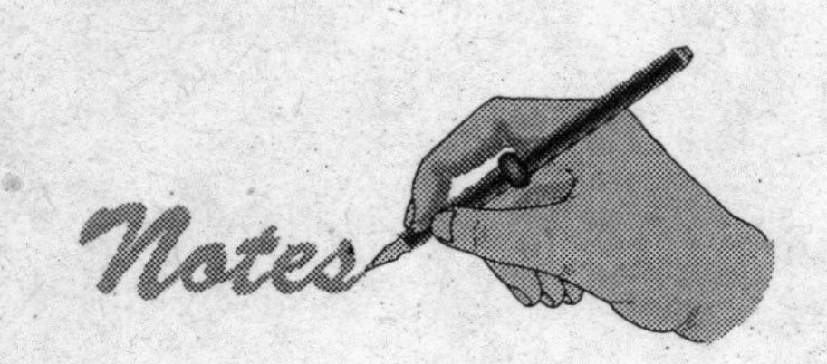

1. This activity is based on the 'disaster crunch model' as proposed by P. Blaikie, T. Cannon, J. Davis and B. Wisner (1994) in: *An Overview of Disaster Management* (1992) UNDP / UNDRO Disaster Management Training Programme.

DISCUSSION MODEL

THE PROGRESSION OF VULNERABILITY

1	2	3		HAZARD
Underlying causes	**Dynamic pressures**	**Unsafe conditions**	DISASTER = VULNERABILITY + HAZARD	**Trigger events**
Poverty Limited access to – power structures – resources Ideologies Economic systems General pre-conditioning factors	Lack of – local institutions – education – training – appropriate skills – local investment – local markets – press freedom Macro-forces – population expansion – urbanization – environomental degradation	Fragile physical environment – dangerous locations – dangerous buildings and infrastructure Fragile local economy – livelihoods at risk – low income levels Public actions		Earthquake High winds Flooding Volcanic eruption Landslide Drought War, civil conflict Technological accident

DISASTER

Which disaster model works for me?

Purpose

This activity aims to clarify the conceptual switch from relief to development. It asks participants to consider how they can get involved in disaster reduction.

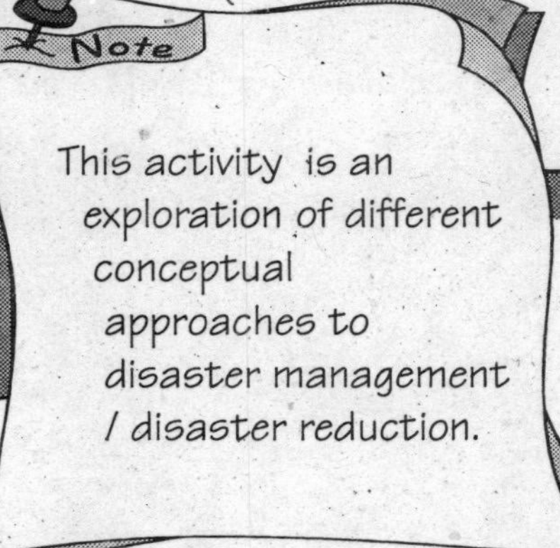

Note
This activity is an exploration of different conceptual approaches to disaster management / disaster reduction.

Procedure

The session involves a number of different activities: input based on various diagramatic representations of the risk - hazard - disaster relationship; creation of a visual presentation of the relationship, and discussion.

Time

- 1½ - 2 hours

Materials

- copies of the slow and rapid onset disaster continuum diagrams for each participant, and on overhead projector slide (see resources)
- copies of the 'Expand-Stretch' model for each participant and on overhead projector slide (see resources)

Process

Introduction

1. Outline the purpose and procedure of this activity.

2. Give a brief input based on the diagrams showing the phases of rapid and slow onset disasters[1]. Explain the different phases (see resources).

3. Ask participants to illustrate each phase with examples of the type of activities normally undertaken in each phase.

4. Initiate a discussion around the merits of this diagram; ask questions such as the following:

   ? Do you agree with the representation of disasters in a continuum of phases?

   ? What are the implications of showing the phases in a cycle?

Participant Action

1. Ask participants to get into groups of four. Distribute flipchart and pens. Give the following instruction:

   — Develop and draw a model or diagram that shows how a progression of increased risk leads to disasters.

   — On the model, pinpoint moments when specific actions aimed at disaster reduction could be introduced.

   Prepare to present and explain it to the group.

   You have 20 minutes to develop the model.

2. Monitor the progress and assist where necessary.

Review and Discussion

1. Facilitate group report-backs: allow each group approximately 5 minutes to present and explain their model; encourage questions of clarification but stall specific contestation and discussion.

Key concepts

Introduction

1. Introduce and explain an alternative diagramatic representation such as the 'contract-expand' model (see resources).

It is called the "contract/expand" model, because it assumes that the disaster management components of disaster prevention, mitigation, response and recovery can be carried out at all the times in a hazard prone community. However, the relative weighting of each component "contracts" or "expands" depending on the relationship between the hazard in the vulnerability of the community.

This model assumes the following:

i. That disasters occur when a hazard exceeds a community capacity to manage it (ie when its vulnerability to the hazard has increased)

ii. That all components of disaster reduction can be carried out concurrently, but with different relative emphases.

iii. That the relative weighting of the activities depend on relationship between the hazard and the vulnerability of the community-at-risk, and the technical or operational mandate of the organisations involved.

Review and Discussion

The aim of this discussion is not so much to develop a 'new', 'correct' model of the risk-hazard-disaster relationship. Rather, participants clarify their understanding of how disaster happens and how it can be prevented, by discussing and arguing different viewpoints.

1. Initiate and facilitate a discussion around the different diagrams:
   - ? What are the advantages / disadvantages of the different diagrams?
   - ? What do they show / fail to show?
   - ? How do they place different emphases on various type of interventions?
   - ? What are the different underlying attitudes to disaster interventions?
2. Ask participants to locate their own practice within the models. How does their work relate to long-term risk reduction interventions?
3. Summarise the session.

1. Figures taken from *An Overview of Disaster Management* (1992) UNDP / UNDRO Disaster Management Training Programme (pp 12-13)

DISCUSSION MODEL ONE

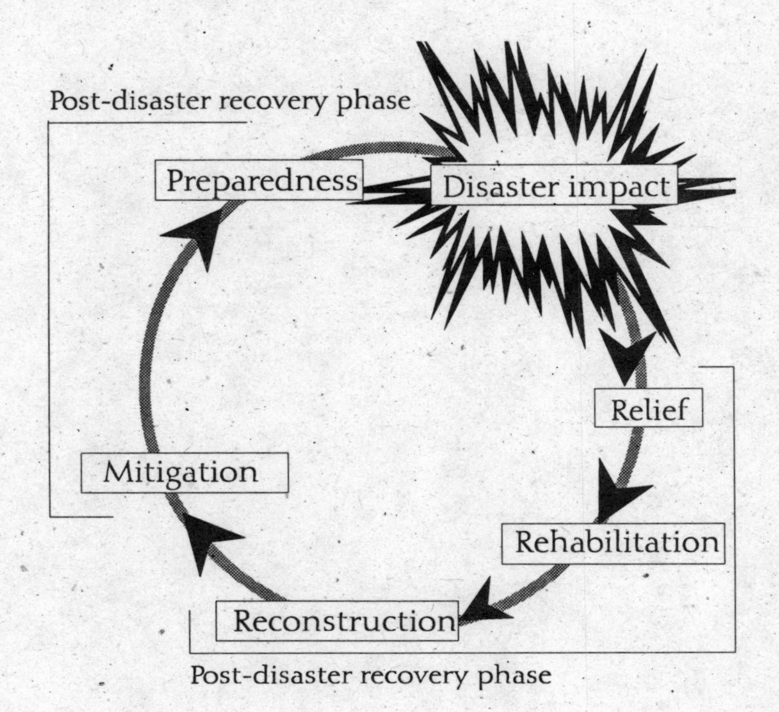

RAPID ONSET DISASTER MANAGEMENT CONTINUUM

SLOW ONSET DISASTER MANAGEMENT CONTINUUM

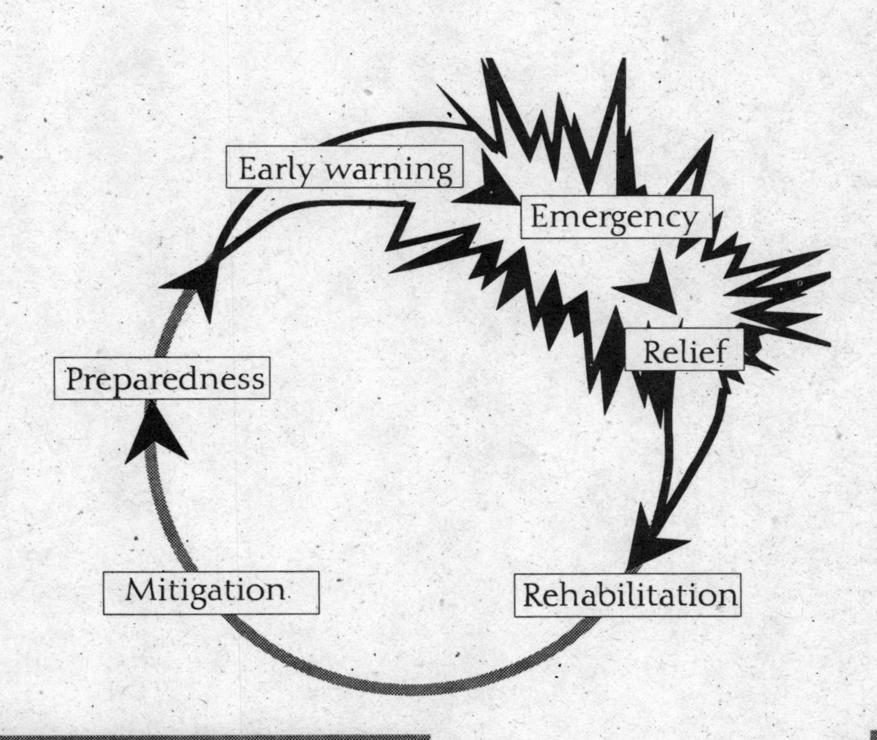

DISCUSSION MODEL TWO

EXPAND-STRETCH MODEL

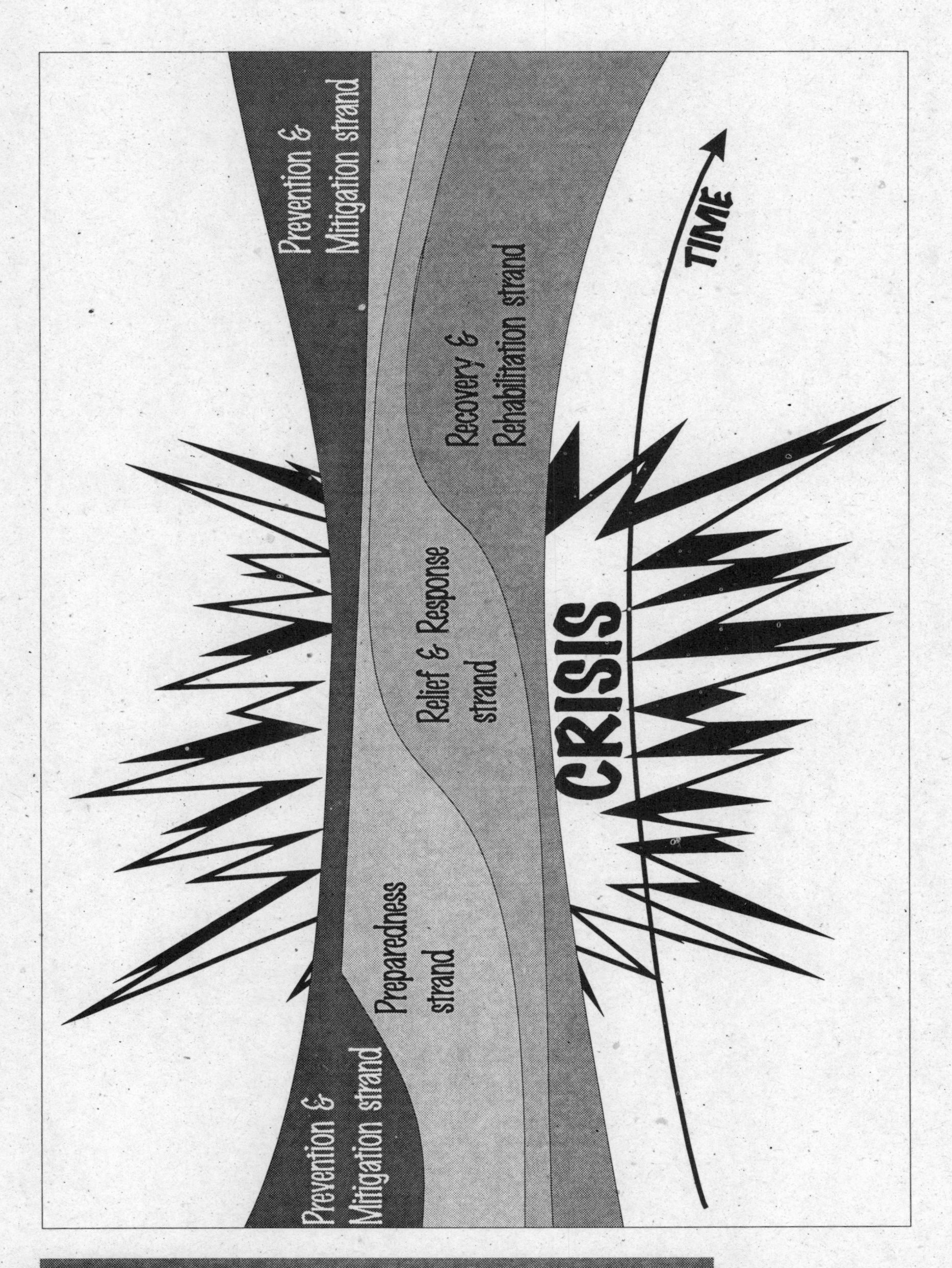

What do we do in emergency preparedness and risk reduction?

Purpose

This activity builds on participants' knowledge of the different categories of disaster management and asks them to classify specific activities in terms of those categories.

Note

The task represents a shift from conceptual to more applied thinking.

This activity is a useful tool for developing participants' awareness of the link between emergency response and risk reduction activities.

Procedure

Participants are asked to classify various activities into their respective disaster management categories. This enables them to analyse actions according to their intended effect within the disaster - development continuum.

Time

- 1 - 1½ hours

Materials

- individual cards describing specific disaster management activities, to be distributed to participants (see resources)
- individual cards with disaster management categories and definitions, to be arranged on the wall or floor (see resources)

Process

Introduction

1. Outline the purpose and procedure of the activity.
2. Distribute cards with disaster management activities amongst the participants.
3. Arrange cards with disaster management categories and definitions as suggested below; either on the floor or on a wall. Ensure ample space to place action cards underneath each term.

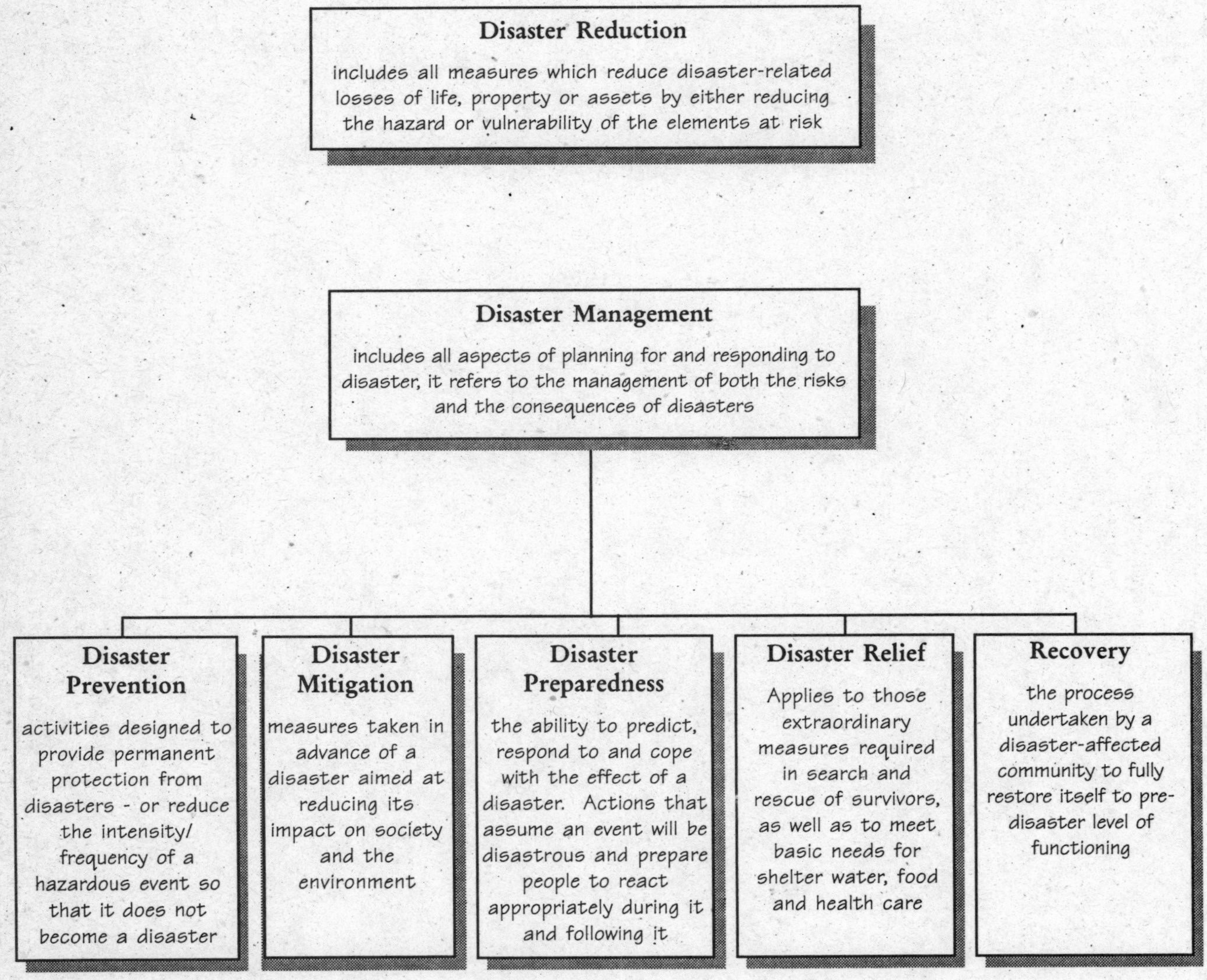

Disaster Reduction

includes all measures which reduce disaster-related losses of life, property or assets by either reducing the hazard or vulnerability of the elements at risk

Disaster Management

includes all aspects of planning for and responding to disaster, it refers to the management of both the risks and the consequences of disasters

Disaster Prevention

activities designed to provide permanent protection from disasters - or reduce the intensity/ frequency of a hazardous event so that it does not become a disaster

Disaster Mitigation

measures taken in advance of a disaster aimed at reducing its impact on society and the environment

Disaster Preparedness

the ability to predict, respond to and cope with the effect of a disaster. Actions that assume an event will be disastrous and prepare people to react appropriately during it and following it

Disaster Relief

Applies to those extraordinary measures required in search and rescue of survivors, as well as to meet basic needs for shelter water, food and health care

Recovery

the process undertaken by a disaster-affected community to fully restore itself to pre-disaster level of functioning

Participant Action

1. Ask participants to consider the activities described on their cards: which category of disaster management do they belong to?

2. Ask participants to place their cards underneath the appropriate category on the disaster management structure when they have made their decision. Some activities might apply to more than one category. If this is the case, a card can be handwritten and placed in the corresponding category.

Review and Discussion

1. When all cards have been placed review the structure, category by category.

   — Request participants to take turns in reading out first the definition of a category, and then the cards placed underneath the category.

   — Ask whether they are satisfied that activities have been placed correctly.

   — Discuss any contested categorising. If no consensus decision can be reached put the card aside for future reference and later consideration.

   — Check participants usage of terminology: do they have a clear understanding of concepts?

   — Encourage any examples that may clarify contested concepts, classification or activities.

2. Point out that some of the terms overlap, for example prevention and mitigation, and rehabilitation and reconstruction.

3. When all activities have been reviewed return to the contested ones and attempt to re-classify them.

4. Allow participants to produce duplicates of any activities that should appear in more than one category. This should encourage discussion around possible overlaps between mitigation and response strategies and activities.

5. Sum up the activity by asking participants to name specific things they have learnt in this process.

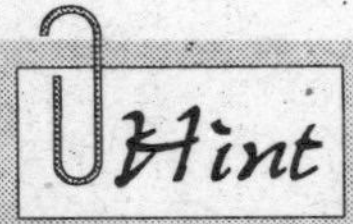

A useful way to follow up this activity is by asking participants to focus on a specific hazard: given this hazard what would be sample activities for each of the categories?

Ask participants to write hazard related activities on cards and categorise them as before.

Suggested Solutions to the Categorisation of Activities

(those cards with an asterisk* could have been duplicated, and appear in more than one category)

Disaster Recovery and Rehabilitation

- Educating demobilised soldiers on their role in non-combat society*
- Seed distribution to food insecure families during drought
- Using drought resistant seed varieties in drought affected areas*
- Food for work projects in a drought (community based)*
- Food for work in a drought (public works schemes)*
- Subsidising the cost of seed and other inputs after a drought
- Drought resistant seeds distribution to food insecure families during drought*
- Reconstructing dwellings following a cyclone
- Reconstructing dwellings with resistant or easily rebuilt materials following a cyclone*
- Restoring community development activities
- Tracing family members separated in a flood or conflict emergency*
- Using flood resistant seed varieties in flood prone areas*
- Establishing community household granaries*
- Providing employment skills to demobilised soldiers

Disaster Mitigation

- Protecting deep and shallow wells in a cholera prone village*
- Planting trees to stabilise a deforested landslide-prone slope
- Conducting household education campaigns on safety with fires before winter months*
- Using drought resistant seed varieties in drought affected areas*
- Food for work projects in a drought (community based)*
- Food for work in a drought (public works schemes)*
- Reconstructing dwellings with resistant or easily rebuilt materials following as cyclone*
- Tracing family members separated in a flood or conflict emergency*
- Using flood resistant seed varieties in flood prone areas*
- Establishing community household granaries*
- Building up earth mounds on which to raise houses above a flood level*
- Drought resistant seeds distribution to food insecure families during drought*

Disaster Prevention

- Achieving 100% measles immunisation coverage in children under five years old following population displacement*
- Protecting deep and shallow wells in a cholera prone village*

Disaster Preparedness

- Intensified training in first aid before multi party elections
- A radio announcement to evacuate from a flood prone area
- Dissemination of information on international humanitarian law to rival military or tribal factions before a political rally
- Training and awareness - building on the risks of dysentery and cholera before the onset of the rainy season
- Tying bells on to ropes strung across a stream or river - up stream to alert people downstream of an impending flash flood
- Rainfall monitoring and reporting
- Prepositioning intravenous fluids, oral rehydration salts and disinfect in cholera prone areas before the rainy season
- Temporary evacuation to primary schools while waters flood subside
- Damage assessment after a violent storm has occurred*
- Drought resistant seeds distribution to food insecure families during drought*
- Food distribution*
- Conducting household education campaign on safety with fibres before winter months*

Disaster Relief

- Damage assessment after a violent storm has occurred*
- Food distribution*

Resources

Disaster Management Activities

Providing employment skills to demobilised soldiers following conflict	Distributing seed to food insecure families during drought
Using drought resistant seed varieties in drought affected areas	Initiating community based food for work projects in a drought
Implementing food for work public works schemes in a drought	Subsidising the cost of seed and other inputs after a drought
Conducting intensified training in first aid before multi party elections	Making a radio announcement regarding evacuation from a flood prone area
Disseminating information on international humanitarian law to rival military or tribal factions before a political rally	Conducting training and awareness - building workshops on: 'the risks of dysentery and cholera' before the onset of the rainy season
Tying bells on to ropes strung across a stream or river - up stream to alert people downstream of an impending flash flood	Monitoring and reporting rainfall
Pre-positioning intravenous fluids, oral rehydration salts and disinfectant in cholera prone areas before the rainy season	Evacuating residents to primary schools while flood waters subside

Disaster Management Activities continued . . .

Doing damage assessment after a violent storm has occurred	Distributing drought resistant seeds to food insecure families during drought
Distributing food	Educating demobilised soldiers on their role in non-combat society
Reconstructing dwellings following a cyclone	Reconstructing dwellings with wind resistant or easily rebuilt materials following a cyclone.
Restoring community development activities	Tracing family members separated in a flood or conflict emergency
Using flood resistant seed varieties in flood prone areas	Establishing community household granaries
Constructing an embankment in a flood prone community	Building up earth mounds on which to raise houses above a flood level
Achieving 100% measles immunisation coverage in children under five years old following population displacement	Protecting deep and shallow wells in a cholera prone village
Planting trees to stabilise a deforested landslide-prone slope	Conducting household education campaigns on safety with fires before winter months

Disaster Management Categories

Disaster Reduction

includes all measures which reduce disaster-related losses of life, property or assets by either reducing the hazard or vulnerability of the elements at risk

Disaster Mitigation

measures taken in advance of a disaster aimed at reducing its impact on society and the environment

Disaster Management

includes all aspects of planning for and responding to disaster, it refers to the management of both the risks and the consequences of disasters

Disaster Relief

applies to those extraordinary measures required in search and rescue of survivors, as well as to meet basic needs for shelter water, food and health care

Disaster Preparedness

the ability to predict, respond to and cope with the effect of a disaster. Actions that assume an event will be disastrous and prepare people to react appropriately during it and following it

Disaster Prevention

activities designed to provide permanent protection from disasters - or reduce the intensity/frequency of a hazardous event so that it does not become a disaster

Recovery & Rehabilitation

the process undertaken by a disaster-affected community to fully restore itself to pre-disaster level of functioning

What is gender?[1]

Purpose

This series of activities aims to clarify participants' understanding of 'gender' as a concept that is crucial to vulnerability assessment and risk reduction.

Note

It is important that participants should not equate 'gender' with 'women'. This activity helps participants to make distinctions between 'gender' and 'women'.

Procedure

This session is composed of a number of different activities: after an input from the facilitator, participants are asked to list stereotypical features of men and women; this is followed by the construction of an 'activity clock' and discussion; finally, groups work with definitions of various gender terms.

Time

- 2 - 2½ hours

Materials

- a sheet of flipchart paper divided into two columns and entitled 'men are...', 'women are...'
- list of 'gender definitions', for each participant (see resources)
- an illustration of the 'I Ching' (optional)

Process

Introduction

1. Outline the purpose and procedure of the activity.

2. Explain that 'gender' is not about conflict, but reconciliation: it is about two halves of a whole and the need for balance between the halves.

   Introduce an illustration of an image or concept that represents the whole, in two evenly balanced halves, such as the 'I Ching'.

In the SADMTP workshop the facilitator explained the 'I Ching' as an image of the energies in Yin and Yang: the eye in each half represents the other in each. She pointed out that this is an image of the universe, and that there is a need for balance in the universe, between opposites that complement each other; she explained the interdependence and reconciliation between the halves.

Participant Action 1

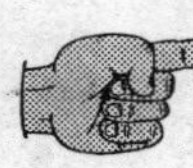

'Gender' is often equated with 'feminists' and facilitators might expect a degree of reluctance, resistance or even hostility to a session entitled 'gender'. In order to diffuse tensions it is advised to begin with a light-hearted investigation of popular stereo-types. This allows all participants to laugh with each other. Throughout the series of activities participants must remain aware that 'gender' refers to both men and women.

1. Divide participants randomly into two groups labelled 'men' and 'women'. Ask groups to think about 'typical features of men / women'.

2. Reveal the prepared flipchart and invite participants to describe men / women. List the adjectives in the appropriate columns.

   (allow approximately 5 minutes for this)

'WOMEN ARE...'	'MEN ARE...'
lazy	selfish
nice and sweet	self-centred
religious	strong
humble	protective
unpredictable	lazy
jealous	productive
weak	close-minded
conservative	stubborn

3. Explain that you will review the lists later.

Participant Action 2

1. Draw the outline of a clock on the flipchart; include the numbers.

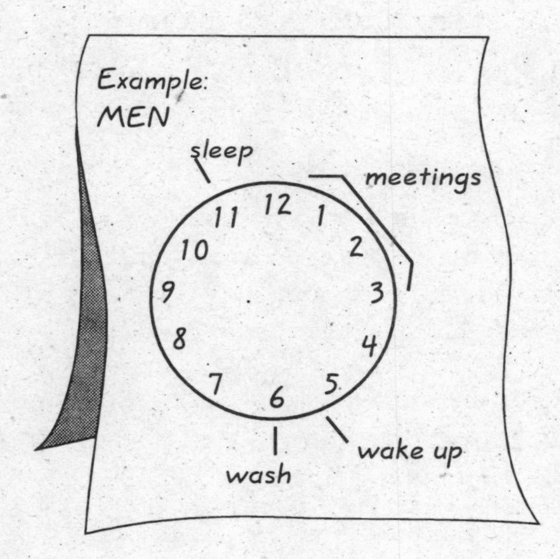

2. Ask participants to get into small groups made up of both men and women. Hand each group 2 sheets of flipchart paper.

3. Ask each group to draw 2 clocks and label one 'men', one 'women' and give the following instruction:

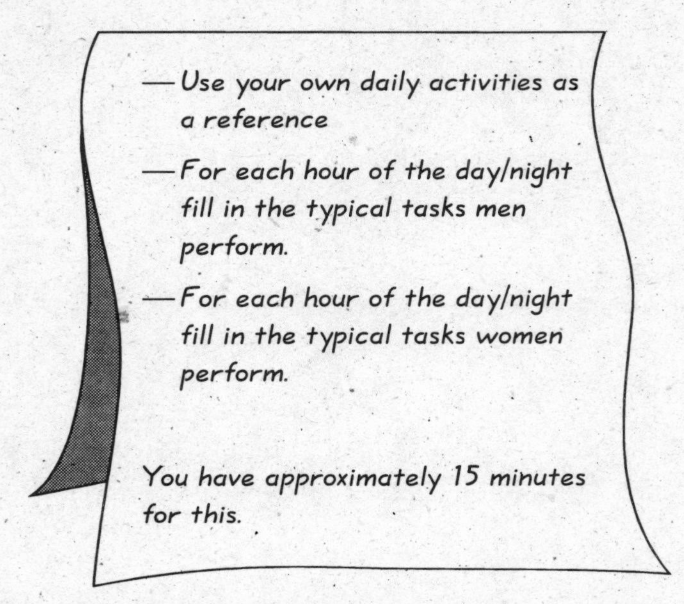

—Use your own daily activities as a reference
—For each hour of the day/night fill in the typical tasks men perform.
—For each hour of the day/night fill in the typical tasks women perform.
You have approximately 15 minutes for this.

4. Ask each group to briefly report back on the activity clocks recorded on their worksheets.

Review and Discussion

1. Suggest that you will now review the lists of 'typical features'.
   - Give examples from the lists to show that both men and women have been described with positive and negative qualities.
   - Give examples of features that are listed in both columns of men and women.
   - Point out that some features refer to the biological differences between men and women; explain that these are differences based on sex, while others refer to the different roles that men and women play in society. This is difference based on gender. Give examples for each, in order to illustrate the difference.

2. Ask participants to refer to their activity clock worksheets and compare the information from the clocks with the information on the newsprint.
   Ask questions such as the following:
   - **?** Are the features listed in keeping with the tasks and activities listed? For example, are women really lazy?
   - **?** Why do some people have the notion that 'women are lazy'?

   Refer to the task clocks and point out those tasks that are 'invisible'; point out how many activities of women are hidden; for example, refer to housekeeping / childcare activities that do not appear in labour statistics.

   - Point out that many of the features listed are stereotypes; ask participants to explore the origin of stereotypes: where do they come from, and how are they disseminated?

3. Compare the type of tasks listed for men and women and conduct a gender assessment:

   — could all the activities be done by either men or women? Why are some activities only done by men / women?

   — point out how men's tasks are often specific, and how one activity often occupies a considerable time-span;

   — point out how women are engaged in a wide variety of activities, and how they switch activities, eg. domestic tasks, income-generating tasks, mothering etc.

   — introduce the concepts of 'productive' and 'reproductive' roles. (see resources for definitions) Ask participants to give examples for both.

   — point out that household tasks such as cleaning, cooking and laundry management are sometimes listed as 'productive' and sometimes as 'reproductive' tasks. Discuss the reason for this categorisation.

   — ask participants to examine the division of labour in their households: who does the work in the home? Who performs income-generating tasks? What is the decision-making procedure over cash and other resources? Is there a balance in the distribution of tasks and responsibilities?

   — introduce the concept of 'community managing roles' (such as serving on committees or attending school meetings). Who participates in community decision-making bodies?

4. Ask participants to do a buzz around the following question:
   - What would make it easier for you to deal with your activities? What would you need to better manage your chores?
   - Note responses on flipchart.
   - Introduce the concept of 'gender needs': explain the difference between 'practical gender needs' and 'strategic gender needs' (see resources for definitions).
   - Ask participants to give examples of practical and strategic gender needs, for both men and women.
5. Suggest a brief break, stretch or energising game.

   (allow approximately 5-10 minutes)

Participant Action 3

1. Affirm how you have just conducted a basic gender analysis.
2. Ask participants to suggest how, in the light of the activities and ensuing discussion, they would define 'gender'.
   - Write up the key terms suggested.
3. Hand each participant a copy of 'gender definitions' and ask them to read and review the definitions, individually.

   Allow approximately 10 minutes for this activity.

Review and Discussion

1. In plenary, encourage questions of clarification. Manage the process, but suggest that rather than you, participants respond to each others' questions. This will allow them to practise using key terms. Intervene when there is a need for information or a fresh perspective.
2. Ask participants to describe how they would use a gender analysis in order to conduct a vulnerability assessment.
3. Ask participants to suggest ways in which gender analysis is a critical part of risk reduction.

1. Much of this session is based on activities facilitated by Paulina Chiziane from the CVM (Mozambique Red Cross) and a participant in the SADMTP course. Her sensitive communication and insightful facilitation resulted in an enjoyable and rewarding session that was highly commended by other participants.

Resources

DISCUSSION DEFINITIONS

Definitions of concepts used in gender analysis[1]

Gender and Sex

Sex identifies the physical, biological difference between men and women. It refers to whether people are born male or female. Gender identifies the social relations between men and women. It refers to the relationship between them, and how it is constructed. It refers to the expectations people have of someone, simply because they are male or female.

Gender Roles

Men and women undertake different roles, depending on the task they are involved in. Gender planning recognises that in most at-risk communities women have a triple role, undertaking reproductive, productive and community managing activities. Men are primarily engaged in productive and community politics roles.

reproductive role - refers to child-bearing/child rearing responsibilities, and to domestic tasks required to guarantee the well-being and health of present and future workforce (other members of the household, including children). Time-consuming activities such as water and fuel collection are examples, as is the gathering of wild fruits during time of drought.

productive role - refers to the work done for pay in cash or kind; these roles include both subsistence/home production with actual use value, and market production with an exchange value. Examples include small income-generating work such as weaving or basket-making, undertaken both for household usage and to decrease economic vulnerability of the household.

community managing role - refers to activities done at the community level, mainly by women, as an extension of their reproductive role. This is voluntary unpaid work aimed at maintaining the well-being of the whole community. For example: work involved in the maintenance of communal water sources, or work in a primary health care programme, or school committee.

community politics role - refers to activities undertaken mainly by men at the community level, often within the framework of policy-making or politics. An example would be the constitution of community representative structures. This work is usually paid, either directly or through status and power.

Gender Needs

Based on their roles women and men have different needs. This is enhanced by the subordinate position of women in Southern Africa.

practical gender needs - are linked to the social roles assigned to women by particular social conventions and culture. Practical gender needs do not challenge gender divisions of labour, but they are a response to immediate perceived necessity. For example, the arduous task of collecting water from remote places in drought-prone areas forces women to walk long distances with a heavy load.

strategic gender needs - strategic needs arise out of the subordinate position of women in society. Meeting such needs would assist women to achieve greater equality and change the existing gendered division of labour. Empowering women through greater self-reliance would be one way of reducing risk.

1. These definitions are adapted from C. Moser and C. Levy (1984-90) *Training Materials for Training in Gender Planning for Development*.

How do economic adjustment policies affect risk?[1]

Purpose

Participants develop their understanding of how economic policies at the national level affect household vulnerability and decrease the capacity of individuals to cope with risk.

Note

Participants enhance their knowledge, test their understanding and use the information gathered to assess economic risk.

Procedure

The session is a series of inputs, discussions and participant actions.

Time

- 3 - 4 hours

Materials

- copies of the SAP play for each participant (see resources)
- a poster / sign saying: TOBACCO MARKET COLLAPSES

Process

Section 1: What is SAP?

Introduction

1. Introduce the session by outlining purpose of the session and procedure of section 1.

2. Begin by giving an interactive input: ask questions, allow participants to respond, and add information and/or explanations, if necessary.

   ? Most Southern African countries are in an economic adjustment process. Why?

   There has been a process of long-term decline or stagnation, due to:
   - economic decline / lack of economic growth
   - drought and other shocks / disasters
   - civil conflcit and population movements (refugees)
   - mismanagement
   - an unfair trading environment (decline in prices for exports)
   - lack of control over decision-making

   ? What are the effects of long-term decline?

   - lower export revenue, and higher import costs
   - a balance of payment crisis (what does that mean? There is less earning than spending; hence there is an increase in borrowing which leads to increased debt; this, in turn, leads to a devaluing of the currency)
   - lower government revenue
   - increasing budget deficits
   - increased interest rates
   - increased inflation

   ? The response to this decline has often been a stabilisation or adjustment policy. How do you define stabilisation and adjustment?

   —Stabilisation:
   refers to economic reforms required by the International Monetary Fund (IMF). It aims at stabilising the economy so that there is a balance between income and expenditure. Stabilisation policies are short-term solutions managed by the IMF.

   —Structural Adjustment:
   is concerned with changing the structure of the economy in order to redress the underlying problems that lead to economic decline. It is controlled by the World Bank (WB) and aims at long-term solutions. It has a sectoral focus, that is, it generally pushes for an increase in the production of goods (industrial or agricultural) that can be traded. It also aims at reducing the size and influence of the public sector.

3. Explain to participants that in order to illustrate structural adjustment you will tell an everyday story.

Sample story (you may want to make up your own):

"When John Kumwembe returned home from years of working in a South African mine he took what money he had and set up a store. It was the only one in the region and although the villagers in surrounding communities were poor his business did well.

One day his friend Kileo came to see him; they used to play in the same soccer team but hadn't met for a while. After the exchange of greetings Kileo explained the reason for his visit: he had bad problems. His wife was ill and needed to go to town to have an operation. His son's schoolfees were due - in short, he needed to borrow some money and would John lend him 1000 kwacha. If the rain came soon his crops would be good, so he could pay the money back.

John agreed: after all, they had grown up together, and Kileo's wife was a good customer at his shop. At the bank the manager warned him; you must charge interest, at least 15%! John gave Kileo the money and said: you must pay me back with interest, so you owe me 1000 plus 15 %, that is 1150. Kileo agreed; he needed the money and besides he knew prices always seemed to go up, why should this be different?"

4. Interrupt the story and ask participants to continue where you left off: what do they think happened next?
   — Allow 5 minutes for participants to spin out the story; collect alternative suggestions without commenting.

5. Continue the story:

" At the end of the harvest season Kileo came to see John. They exchanged greetings and news: the wife was well, the operation had been a success. His son was doing well at school. For John, business was not too good - people were poor and were doing without some of the things they used to buy. Yes, said Kileo, these were bad times. The rains didn't come and his crops had died. So he had nothing to sell - in fact, they wouldn't even have enough to live on. And now, there had been cut-backs at his work, and he had been laid off. But the school fees were due again. Please could John lend him some more money - 500 kwacha; he would owe him 1650.

John wanted to help him; after all, he was a good man and he worked hard to support his family. And their children went to the same school. But how would Kileo ever be able to pay him back? He spoke to the bank manager and when he came back to Kileo he said: "I will lend you some more money, but this time there have to be some conditions."

6. Interrupt the story again; ask participants: what conditions would you suggest should be imposed on Kileo?

   — Allow 5 minutes for participants to make suggestions for conditions attached to the loan.

7. Continue the story:

" John told Kileo that he would have to change his way of life: if he could not afford to pay school fees he would have to take his child out of the school; he should grow different crops; use his fields to grow something he could sell, like tobacco. How could he afford to buy the seeds? Well, he might have to sell some of his cattle. And he would have to begin making payments every months. He, John, could no longer wait until the end of the season. And he was sorry, but the interst had gone up. 16% now. Maybe his wife could go out and work and help with the repayments?

Kileo went home that day, and his life and that of his family changed forever."

A year passes: Hold up the sign that reads: *Tobacco Market Collapses...*

8. Ask participants to describe briefly, how this story illustrates 'structural adjustment':

   — at the household, community, national level

   — in terms of the relationship between the IMF/World Bank and a country

   — in terms of the impact of hazards such as drought on 'normal' development processes

9. Suggest a break / energising game.

Section 2: The aims of SAP

Introduction

1. Introduce the procedure of this section.
2. Ask participants to focus on the characteristics of structural adjustment policies. Ask: How do they work? What are the goals of SAP?
3. Record responses on newsprint; ask participants to explain to each other what each one means and assist where necessary.

   The following main points should be mentioned:

   —increase exports and reduce imports, to assist balance of payments difficulties and enable Southern African countries to

   — repay western debts

   — increase the production of food and cash crops

   — encourage private-sector involvement and competition

   — reduce government expenditure and the role of government intervention: liberalisation of trade and industry

   — economic growth, efficiency, free markets

Participant Action

1. Introduce 'The Structural Adjustment Play' (see resources) by outlining its purpose to present arguments for and against the introduction of an economic adjustment policy, as a way of exploring whether / how economic policies can reduce or enhance risks
2. *The Structural Adjustment Play*
   - Ask two participants to act as 'supporting voice' and 'opposition voice'
   - Give each acting participant a role sheet
   - Point out that you will act as the 'presenter'

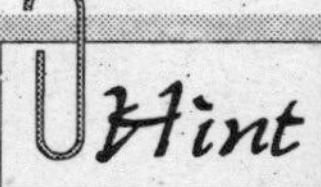

Alternatively, you can ask participants to improvise a dialogue on the pros and cons of SAP. This could either be in the form of a panel discussion, with approximately 3 participants on either 'side'; or in small 3-member groups

3. Manage the process and participate in the reading / enactment of the play

4. Encourage any questions of clarification; explain terms and concepts, where necessary.

5. Ask participants to summarise the main arguments made by the opposing voices; suggest that you will record them as a cost-benefit analysis:

   — Prepare a sheet of newsprint in two columns, entitled: 'benefit: who has gained?' and 'cost: who has lost?';

   — Record points raised in the appropriate column.

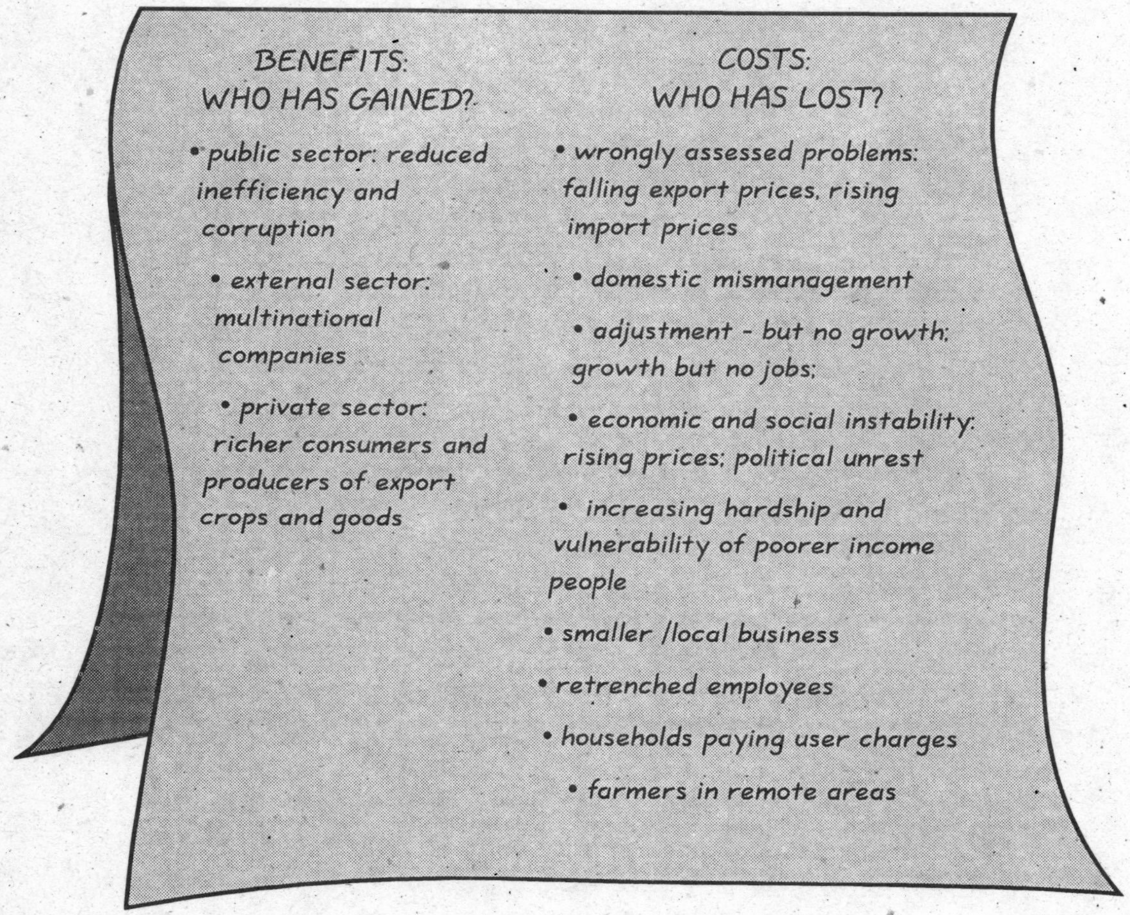

BENEFITS:
WHO HAS GAINED?

- public sector: reduced inefficiency and corruption
- external sector: multinational companies
- private sector: richer consumers and producers of export crops and goods

COSTS:
WHO HAS LOST?

- wrongly assessed problems: falling export prices, rising import prices
- domestic mismanagement
- adjustment - but no growth; growth but no jobs;
- economic and social instability: rising prices; political unrest
- increasing hardship and vulnerability of poorer income people
- smaller /local business
- retrenched employees
- households paying user charges
- farmers in remote areas

6. Ask participants to respond to the following question:

   ? From your experience, which voice (SV or OV) or which column (benefit or cost) reflects the impact that an economic adjustment policy has had in your country? Explain.

7. Suggest a break and/or energising game.

Section 3: How have structural adjustment policies increased risk?

Introduction

1. Outline the procedure of this section.
2. Divide participants into five small groups with the following labels:
   - — social impact
   - — health impact
   - — environmental impact
   - — political impact
   - — women
3. Give the following instruction:

List ways in which SAP have increased risk with regard to:
— social impact
— health impact
— environmental impact
— political impact
— gender isolation
You are expected to report back to the rest of the group.

You have 20 minutes to complete the task.

Review and Discussion

1. Manage report backs, and facilitate discussion.
2. Sum up increased vulnerabilities; ask participants to name some of the capacities which communities have relied on in order to cope with increased threats.

1. Much of the information is based on training sessions prepared and run by Carol Thompson, University of Arizona, and Graham Eele, from the Southern African Development Community Food Security Training Programme, Harare.

Notes

Resources

PLAY

Discussing structural adjustment

This is the skeleton of a possible dialogue on structural adjustment policies (SAP) between a supporting voice (SV) and an opposition voice (OV). Arguments can be elaborated at will; 'actors' should feel free to enact the moods underlying the arguments. Each characteristic of SAP will be introduced briefly by the presenter (P), after which SV and OV will engage in a dialogue.

P: Ladies and gentlemen, welcome to our dialogue on SAP. May I introduce to you the two contestants: one my right: SV, and on my left: OV. They will present to you arguments in support of and in opposition to SAP - after which you will be invited to make up your own mind! (applause)

P: The first (but not foremost) characteristic of SAP is the liberalisation of trade and exchange. The government no longer controls the price of goods sold in a country.

OV: Yes, we have noticed that! Look at how the prices have rocketed! The average household can no longer afford to buy basics like sugar and chicken.

SV: We no longer have to endure the tyranny of Government control! We, the business owners, will at last have a say over what to produce, and what to charge for our products. Our profits are your benefit: we can expand our business, and that means you have a much greater choice of goods! More variety! And: more jobs - the more business, the more employment. If we can show the world that we have a thriving economy, foreign investors will be confident to invest in our country - and their companies and factories will mean more jobs.

OV: Foreign countries already own too much of our industry. They only have their profits at heart - not the benefit of our country. We will lose control of our economy and resources.

P: Trade liberalisation also means, wages are no longer controlled by the government. Collective bargaining is the order of the day: workers negotiate directly with their employers and agree on the rates of pay.

SV: We want to ensure high productivity - so we will pay good wages. Of course, we can only pay what we can afford in terms of ensuring growth. Remember: increased productivity - increased goods, and increased jobs!

OV: Wages are already so low that most of the workers cannot afford to purchase a basic food basket for their households. With this rate of unemployment we are not in a position to bargain with employers. They will just dismiss us! If the government doesn't control our wages and rights through labour laws, we will have to accept even less from employers. 'Freeing wages' is a euphemism for 'banning trade unions'.

OV: You want us to grow crops that can be sold. But who will travel into the remote parts of the country and pick up our products? Who will pay the additional cost of transport? And what if the market crashes: if our maize does no longer fetch a good price, internationally?

We have to ensure that we have basic food-stuffs at home, before growing for export. We should grow more drought resistant crops like sorghum and cow-peas, but who will be prepared to grow food if high incentive prices are offered to grow maize?

P: Under the old laws foreign investors could only send a small proportion of their profits out of the country. With SAP they are free to remit all their profits.

SV: At last, a truly free market. This makes it so much more attractive to invest in our country! Why would they want to bring their money here, and provide jobs for you and then not be able to take their profits out? We need foreign currency - how else will we compete in the market?

OV: Can you not see that already too much money is leaving our country? Foreign companies use our resources to get rich - and they leave us all the poorer. Yes, we need foreign currency - but if they are allowed to take out more money we will have an even greater shortage of foreign currency.

Reduced public spending is one of the cornerstones of SAP. This includes cutbacks in money spent on services such as health and education; dismissal of public servants; reorganisation of inefficient systems of management.

SV: For too long too many public servants have been unproductive. To employ three people doing the work of one person is not cost-effective.

OV: So: more unemployed join the queue outside the labour bureau.

SV: The money saved on redundant civil servant's wages can be used much more productively. It means we have a bigger budget for roads, bridges, sports stadiums and the like.

OV: You want to cut on basic services! We fought our liberation for the right to have primary health care, and primary education for all. If people have to pay in order to attend the clinic they will not be able to afford to come. Disease will increase.

SV: We will ensure continued PHC and education - with the support of the World Bank. And we will subsidise basic food stuffs.

OV: What will you do about the spiralling interest rates? Small business sector cannot afford them - but the international companies can go elsewhere for credit.

P: Can we ask both voices to present closing arguments, please.

SV: There has to be growth: no jobs, no income, no stability. There is a lot of evidence of improved economic performance in African countries due to sap's. The annual growth of goods and services in countries like Burundi, Gambia, Kenya and Senegal rose from less than 2% before SAP to 3%. If it wasn't for the disruption by political changes in some countries it could have gone even higher. Besides, the SAP provide an effective instrument to deal with the debt problem. And now, the IMF supports new and more concessional approaches by creditors.

OV: SAP makes too many assumptions: for example, that an improvement in the national economy will ensure an improvement at the household level (not just of the well-off minority); or: that we have a stable political situation, rather than constant upheaval where the changes in one country potentially affect the whole region. And what about the droughts? Women already supply so much free labour to collect water and fuel, and many men and women lose time to productive activities, because of relief work. Times of large-scale food insecurity are not the right time for introducing structural adjustment policies!

What are the causes of droughts and how do they impact on the environment?

Purpose

Participants enhance their understanding of the causes and effects of droughts.

Participants gather technical information on and improve their understanding of

- different types of drought,
- factors effecting the severity of drought, and
- the impact of drought on various elements at risk

Procedure

This is a reading and discussion activity.

Time

- 1½ -2 hours

Materials

- reading: Drought Disaster hits Zimbabwe (see resources)
- copies of 'The Climate Factor' from: *State of the Environment in Southern Africa*, for each participant; (see resources)
- questions written up on flipchart paper
- four sheets of flipchart paper entitled:
  - Identify the elements at risk from a drought
  - Identify human activity that exacerbates the effects of a drought
  - Identify various pressures on the environment as a result of a drought
  - Identify the impact on industry

Process

Introduction

1. Outline the purpose and procedure of the activity.

2. Tell a brief story that illustrates different perceptions of causes and effects and definitions of what might constitute a drought.

> ***Sample Story***
>
> *A simple story may begin with a letter from an African person writing from northern England. 'We have water restrictions', she reports, 'because it hasn't rained for three weeks. There is a drought'. The response at home is a mixture of hilarity and disbelief: 'Three weeks no rain and they call it a drought! Ha ha, it always pours in England, it's as green and luscious as can be - they don't know what it means to have a drought!'*
>
> *And yet northern England is experiencing a drought, because 'drought is a relative, rather than an absolute, condition'.*

3. Inform participants that in order to better understand drought as 'a relative rather than absolute condition' you will hand out some technical information, written about the Southern African region.

Participant Action

1. Distribute copies of 'The Climate Factor' and ask participants to read through the information, underlining or writing down key words and issues.

   Allow approximately 30 minutes for this.

Alternatively, you may invite a technical resource person who has expert knowledge of droughts to give a presentation on the topics outlined. In this case ensure that participants end up with written records of key issues.

Review and Discussion

1. Ask participants to work in pairs and to refer to the reading when following the instructions written on newsprint:

   Face each other and take turns in explaining the following ideas and concepts:

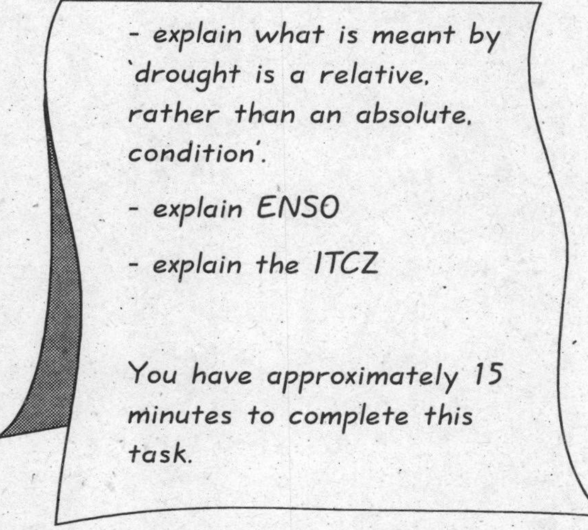

- explain what is meant by 'drought is a relative, rather than an absolute, condition'.
- explain ENSO
- explain the ITCZ
You have approximately 15 minutes to complete this task.

2. Monitor the process and assist where necessary.

3. In plenary, check how participants found the task and encourage questions of clarification and further explanation.

4. Ask participants to consider the sources of information that technical advisers draw on: list the sources for meteorologists, hydrologists and agriculturalists on flipchart.

Participant Action

1. Stick the four sheets of prepared flipchart paper on the wall:

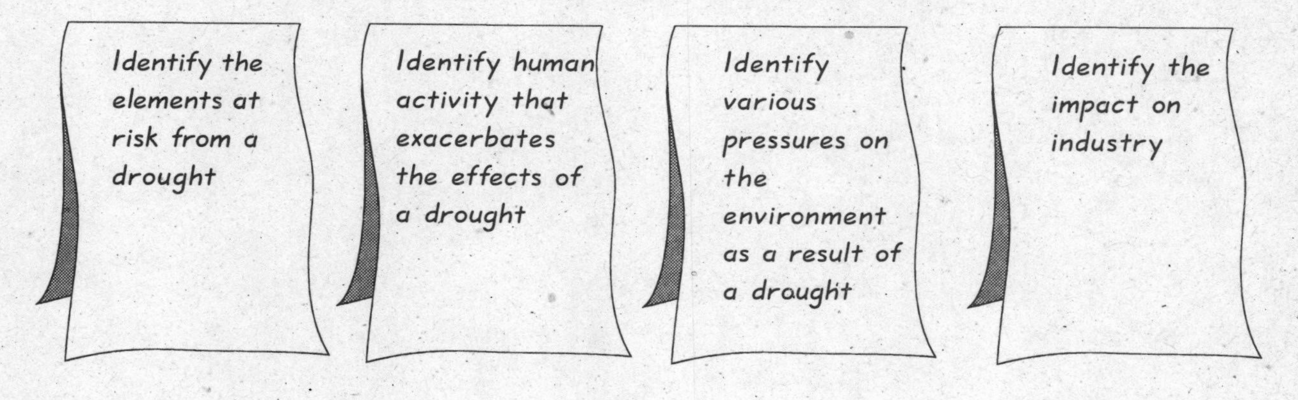

Identify the elements at risk from a drought
Identify human activity that exacerbates the effects of a drought
Identify various pressures on the environment as a result of a drought
Identify the impact on industry

2. Ask participants to get into four groups and ask each to choose one flipchart. Request groups to respond to the instruction on their flipchart. Allow approximately 5-10 minutes.

3. Instruct participants to go on a 'gallery walk' and read the lists of other groups.

Review and Discussion

1. Review the lists: highlight important aspects and add other factors that may have been omitted.

2. Ask participants to refer to their experience of a recent drought, such as the Southern African drought of 1991-2.

   Discuss the following questions:

   ? What impact did the drought have on you, personally?

   ? What impact did it have on other people and communities in your area / region / country?

   ? List social, organisational, economic and environmental factors that contributed to your being able to cope with the drought.

   ? Identify factors that made others more vulnerable.

3. Summarise the activity by asking participants to name ways in which they could apply the information gathered to their work in the field.

4. Distribute copies of 'Drought disaster hits Zimbabwe' and ask participants to read this text as follow-up homework.

Resources

READING

DROUGHT DISASTER HITS ZIMBABWE

By Joshua Chigodora and Munyaradzi Chenje

(Environment Resource Centre for Southern Africa; SARDC)

Maria Kudzingwa vividly remembers the years of plenty in Nhekairo village, in Wedza. The rains were more reliable and rivers in her area flowed swiftly. All those memories seem, however, to be drying up like the thirsty, cracked earth on which she is sitting, patiently waiting at the borehole for her turn to fill her water pail.

With only six people at the borehole, including Kudzingwa — who is in her sixties — the atmosphere appears laid-back, belying the jostling and elbowing for position which usually takes place on a busy day. Kudzingwa sits close by to rest her legs tired from the 10-kilometre walk from her home to the borehole. She turns her wizened face to watch those ahead of her fill up their containers. Up and down! Up and down, the handle of the hand pump goes, pumping water from the bosom of the earth. With each upward motion, the water flows into the pipe and is spat out, gushing into containers with each downward motion. Some of the water misses a container, splashing by its sides and trickles down the sloppy ground. Kudzingwa's eyes seem to follow the trickle of water as it meanders its way on the ground. The flow gets slower and slower as the thirsty earth sucks up every drop of the water.

Nhekairo village, in Wedza, about 150 kilometres east of Harare, used to have abundant supplies of water, says Kudzingwa, finally turning up her face to her visitors. The village used to have plenty of water trees, such as the mukute trees but these have since been cut down, leaving the terrain bare and degraded. For Kudzingwa, the past holds the best of her life experiences. The present, with its severe droughts and failed crops almost every season, is a stark reminder of paradise lost.

Many of the villagers hardly harvested anything this past season as most of the crops wilted under the merciless heat from the sun and lack of rain. While the villagers were self-sufficient in food in the past, most of them have been forced to depend on handouts - government drought relief and welfare. "I should be busy preparing my maize harvest at this time of the year, but here I am busy looking for water," said Kudzingwa with disappointment in her voice.

She points to a place where the village used to draw water from a spring which had water all-year round. Today, the same spring is only remembered in the past tense. Villagers also depended on hand-dug, shallow wells. But those have since given way to drilling rigs which are now used to tap an eve-receding water table deep down in the ground. Kudzingwa blames modern technology for such misfortunes, saying people no longer respect their ancestors and culture. She said the practice of praying for and receiving abundant rains has died and will never return as the spirits are upset with what is happening in this country.

As generations come and go, memories of springs with abundant water supplies, many hand-dug wells, and rain-praying ceremonies would probably only be remembered in folklore. But that too may be difficult to sustain for long, as the present generation seems to have been caught in vicious circle recurrent drought, large-scale crop failures, and the scorching sun. Simba Nherera (28) laments about sharing water supplies with livestock. Water scarcity in the village is evidenced by Manyimo dam which is drying up. Due to lack of many water sources, the villagers now use water from the dam for washing, drinking, cooking and livestock. The water is polluted and looks unfit for human consumption.

The plight of Nhekairo village is not an isolated freak of nature, but is replicated hundreds of times throughout Zimbabwe, with millions of villagers now facing not only starvation but also water shortages. In Tsholotsho, Matabeleland North Province, a total of 35 children died between January-June this year due to malnutrition. In the same period, the district hospital attended to 119 children who were suffering from malnutrition. In Mt Darwin, some villagers are being forced to walk 20 kilometres in search of water. It has been reported that the water table has gone down and boreholes are drying up. In Chivi South in Masvingo province, villagers continue to live on the threshold of starvation.

The water crisis extends beyond rural areas and have forced many urban areas, including Harare and Bulawayo, to start water rationing. Government officials say the present water situation is critical in most towns. In Chegutu, a critical water shortage is threatening industry and some business people fear they could be forced to stop production.

A few months ago the government declared the town, which is more than 80 km west of Harare, a water shortage area and the local town council was forced to introduce water rationing to conserve water. The town's water supply will only last until October this year despite the water rationing. Apart from the fears of economic ruin if water runs out, residents are also concerned about the health risks. Because water is available at specific periods, sanitation and personal hygiene is suffering, according to some residents.

The District Environmental Health Inspector at Chegutu District Hospital believes the possibility of an outbreak of an epidemic should not be ruled out. "The restrictions in water use, though necessary, will result in a decline in levels of personal hygiene. Sewerage blockages are inevitable," said the inspector. "The frequent water cuts will affect the effectiveness of chlorine and therefore the water quality. What do you expect from such a situation?" "The truth is we have had more dysentery cases over recent weeks than at any time before water rationing ...," says Chegutu District Nursing Officer, Sister Patricia Zari. "Nutritional levels have gone down considerably in some areas since we had to close our nutritional gardens because of lack of water"

The water crisis has also affected industries in the town and the surrounding farms. Production in most firms has gone down by 30 percent and the business community fears this could have repercussions to employment and income levels, and the future development of the town. Agricultural activities have already suffered from the water cuts as farmers have been ordered to limit water use to the barest minimum.

The Zimbabwe government has since described the crisis as a disaster, which, if not dealt with promptly, could see many people dying of hunger. The 1995/96 drought is already described as worse than the drought of 1991/92, then called the worst this century. (SARDC)

5 The Climate Factor

Rainfall is the lifeblood of southern Africa. Much of the region is arid or semi-arid and rainfall is extremely variable, often unreliable. A poor year can result in large-scale crop failure, food shortages and, in extreme cases, famine. Trees and grasses wilt and die, and animals perish from hunger and thirst. Subsistence farming, which provides most people of the region with their food, depends on sufficient rainfall. Any signs of drought are received with dread. Drought is associated with suffering, and loss of valued crops, livestock and wildlife. Praying for rain is not uncommon in many parts of the region, and the onset of the rains is often viewed as the single most important event of the year.

Droughts are not easily predicted and most of the region has long, dry spells every year. If the rains do not come by a particular time, this may be a sign of drought; early rains may hide an impending drought.

The 1980s saw an increase in the severity of the impacts of droughts. The 1982-83 drought in Mozambique was considered the worst in 50 years and led to many thousands of deaths.[1] But the worst drought in the region was experienced a decade later, in the 1991-92 season. The areas surrounding the Kgalagadi — the savannas of Angola and Botswana, and the pastoral areas of southern Africa — were the most affected. Drought left most of those areas parched and gasping for elusive rains. [2]

DROUGHT DEFINED

The word "drought" is derived from an Anglo-Saxon word *drugoth*, which means dry ground. However, meteorologists, hydrologists, agriculturalists and economists define drought differently.[3] Meteorologists define drought solely on the basis of the degree of dryness and the duration of the dry period. Hydrologists link periods of shortfall to the effect on surface or sub-surface water supply (stream flow, reservoir and lake levels, groundwater). Agriculturalists link drought to agricultural impacts, focusing on precipitation shortages, the differences between actual and potential evaporation, and factors such as soil water deficit. Economists associate the supply and demand of goods and services with elements of meteorological, hydrological and agricultural drought.[4]

Central to all definitions is the shortage of water. Drought denotes dryness and should not be confused with aridity even though both are characterised by a lack of water. Aridity is a permanent climatic condition. Drought is temporary. Drought occurs when there is a protracted shortage of water. A dry spell has to last long enough to cause damage, otherwise it is not a drought.

Drought is a relative, rather than an absolute, condition. An area normally receiving 1,000 millimetres (mm) of rain annually would experience drought if it received 700 mm per year, but would not have serious problems from an agricultural perspective. An area normally receiving 700 mm would experience crop losses if it received 400 mm. Yet each area would have lost 300 mm. In drier areas, even small reductions of rainfall can have significant economic effects. Although we tend to associate droughts with areas of low rainfall, they can also occur in areas that normally enjoy abundant rainfall.

Shortage of rainfall is always the "trigger", but it is the lack of water in the soil, rivers or reservoirs which causes the hazard.[5] Plants do not use rain as it falls, but rely on the moisture available after the rain seeps into the ground. Similarly, rainfall does not supply water directly for irrigation or domestic use. It is harnessed from rivers and reservoirs, and from underground.

When the duration of a drought is short, the impact is minimal as long as the previous season was wet enough to provide adequate food and water reserves. If a dry spell lasts for a longer period of two or three years, then the impact can be severe on the environment, people and their crops, and livestock. Meteorologists refer to the frequent but short dry-periods, often lasting for less than a year, as "normal drought". It is not uncommon for a "normal drought" to occur in the middle of a wet spell.

CLIMATIC CAUSES OF DROUGHT

Although climatologists have produced a number of plausible explanations as to why droughts occur, a single conclusive answer is yet to be found. Past occurrences of drought have been linked to certain events such as *El Nino* and volcanic eruptions.

El Nino and the Southern Oscillation (ENSO)

El Nino is a weather condition which begins with the warming of waters in the western Pacific ocean, eventually affecting global climate.[6] This condition has an effect on weather over a quarter of the world's surface. *El Ninos* develop as the warm waters of the tropical Pacific spread eastward in concert with shifting patterns of atmospheric pressure. These natural warming events alter weather patterns worldwide, probably causing droughts in southern Africa or contributing to their severity.

During the 1982-83 season, the severe drought in southern Africa and the Sahel, and the famine in Ethiopia, were linked to an *El Nino* occurrence.[7] Again, when the devastating 1991-92 drought occurred, *El Nino* lasted until the end of February 1992.[8] Some experts say that about one-third of the droughts in the region could be attributed to *El Nino.*[9]

El Nino is a component of another global weather phenomenon, the Southern Oscillation, and together these are known as ENSO. During an ENSO phase, equatorial waters across the Pacific ocean get warmer. Normal airflow moves westward from the Pacific to the Indian ocean, but during *El Nino* this movement is weakened or altered. This results in high rainfall in some parts of Latin America but low rainfall and even drought in southern Africa.

The opposite extreme of the ENSO cycle occurs when a cold phase known as *La Nina* or (anti-*El Nino)* is experienced. The occurrence of *La Nina* results in unusually heavy rain in southern Africa. At this time the Pacific is cooler than the Indian ocean and wind moves from the Pacific toward the latter. *El Nino* means "the boy-child" in Spanish, so-named because it occurs around late December, when Christians are celebrating the birth of Christ. La Nina means "the little girl".

Volcanic eruptions

Volcanic eruptions elsewhere in the world have been linked to drought in southern Africa. Climatologists believe that the eruption of Mount Pinatubo in the Philippines in June 1991 could also be linked to the drought that devastated southern Africa in 1991-92.

Dust spewed by the volcano could have interfered with southern Africa's Intertropical Convergence Zone (ITCZ), which brings rain to much of the region, and other weather systems. Mount Pinatubo's volcanic dust reached the stratosphere (a layer of the atmosphere 10-60 km above the earth's surface) over the Indian ocean and partially blocked the sun's radiation. As a result, the ocean and the air above it did not warm as much as is usual. Thus, the rain-bearing wind, which drives moist air toward the region from the north-east, was not strong enough to reach southern Africa and was pushed up north of Zimbabwe.[10] Rain associated with the ITCZ fell further north than usual — over Zambia.

Global climate change

Scientists predict that global atmospheric changes could disrupt established weather patterns, so that existing weather conditions such as drought occur more frequently. It has been suggested that global warming may have caused or contributed to recent droughts, but there is no scientific evidence yet to support this. The fact that droughts have been an ongoing occurrence since pre-historic times makes it difficult to assess the role of global atmospheric change.

SOUTHERN AFRICA AS A DROUGHT-PRONE REGION

Southern Africa's climate and rainfall patterns have been highly variable for at least the last three centuries,[11] leading to recurrent droughts of varying severity.

Droughts lasting between one and five years may occur in isolated areas or on a regional scale.[12] The region experiences regular wet and dry spells, that is, several years of abundant rain followed by periods of little rain. Theories about the cyclic nature of rainfall in the region were first put forward by scientists in 1888. By 1908, a South African scientist based in

Natal had found evidence of an 18-year cycle of wet and dry years. Continuing research seems to support this theory.[13]

A 77-year study done in the upper Vaal catchment, South Africa, reveals two distinct patterns of rainfall cycles. First, the cycles of drought followed by high rainfall are a common feature in the Vaal catchment. Second, the length of droughts in the area varies considerably, while there is also variance in the quantity of rainfall received in each wet period. South African experts believe that this has some significance for the region. If the trend continues, the rest of the 1990s could be relatively wet.[14]

Generally, arid and semi-arid areas are most prone to drought, especially toward the southwest, in Botswana and Namibia. The erratic and unpredictable rains in Botswana make that country more susceptible to water shortages than some of its neighbours. The poor sandy soils, which retain little water, coupled with high evaporation rates, exacerbate the problem. However, a plant that can root deep into the ground has access to moisture for a longer period. The camel thorn and Zambezi teak are good examples. In Botswana, only three years between 1959-1972 had abundant rains; six received inadequate rain and the other five received very little rainfall.[15] While Botswana has recently received above-average rainfall on a national scale, the distribution remains erratic and there are large variations, which cause severe localised crop failure.

Like Botswana, Namibia is an arid country. The northern parts and coastal areas, in particular, are the most severely affected by drought. Parts of the Namib desert record virtually no rain in some years. Evidence of drought and degraded land is particularly strong across northern Namibia. Years of abnormally low rainfall and overuse have turned savanna grass to straw, and dried up creeks and water holes, leaving cattle thin. Dust devils dance over the plains, and sudden bursts of wind can bring a blizzard of white dust, dense as mist.

Photo 5.1

SOUTHLIGHT-G Tillim

Drought often means a daily trek for water, several kilometres longer than usual, as local supplies dry up, shown here in Lesotho.

David Livingstone's journal

Box 5.1

British missionary David Livingstone recorded a drought in Botswana between 1845-1851.

Livingstone was living among the Bakwena at Kolobeng, not far from Gaborone. His letters and diaries described a number of events related to the drought, including the summoning of rain-makers. So bad was the drought that some people had to emigrate. In January 1849, Livingstone wrote that the famine was so acute that the people had survived the previous six months entirely on locusts.

Ultimately, the local people began to blame Livingstone's mission for the drought. The criticism became more vocal when the Kwena chief, a renowned rainmaker, converted to Christianity and gave up rainmaking.

"We know a great difference in him since we came to him. He was a rain maker and had the reputation of being a wizard .... He has nothing now to do with the rain making incantations and it was by his own desire that we began prayer meetings in his house."

(Letter from David Livingstone in Kolobeng to Dr. J.R. Bennett, 23 June, 1848)

Soon after Livingstone's departure from the area, the drought broke.

SOURCES: Hitchcock, R.K., "The Traditional Response to Drought in Botswana", in *Symposium on Drought in Botswana*, Botswana Society, Gaborone, 1978, p. 92
Holmes, Timothy (ed.), *David Livingstone Letters and Documents 1841-1872 - The Zambian Collection at the Livingstone Museum*, The Livingstone Museum in Association with Multimedia Zambia, Lusaka, and James Currey, London, 1990, p. 31

Photo 5.2 NAMIBIA REVIEW-National Parks & Wildlife

Namibia's Etosha pan, now a dried-up lake, supports a large population of wild animals, with ample waterholes in the wet season.

One of the areas that has been severely impacted by drought is Namibia's Etosha Pan, which has attracted the attention of US scientists from the University of Virginia. They are probing air and soil around the pan, now a dried-up lake. In wetter years, it was one of Africa's major breeding grounds for flamingos. When the lake last held ample water, in 1979, more than 200,000 flamingos nested on its islands.

Even in Tanzania, which gets more reliable rains than elsewhere in the region, rainfall can be low and uncertain in parts of the country. For example, in the capital city of Dodoma, in the central part of the country, the rainfall is extremely variable.[16] Tanzania experienced droughts from 1943 to 1993. Recent research in central Tanzania indicates a serious drought and famine in every 10 years, characterised by years ending with the number 3 or 4 in every decade.[17]

Trends and patterns in regional rainfall

Southern Africa's rainfall is variable, but there has been no evidence of long-term change in rainfall patterns. Nor is there any indication that the region's climate is becoming drier overall.[18] A 100-year study (1880-1980) of weather patterns in East Africa — based on air temperature, rainfall, mountain glaciers and lake levels — detected no long-term climatic change.[19] Only short-term climatic fluctuations were noted. Similar studies of long-term records have been car-

Climatic change 1800-1992

Table 5.1

A historical overview of drought and rainfall patterns in southern Africa since 1800.

- ❖1800-30 — Southern African rivers, swamps and other water sources dried up. Some well-watered plains turned to semi-arid karoo.
- ❖1820-30 — This was a decade of severe drought throughout Africa.
- ❖1844-49 — Southern Africa experienced five consecutive drought years.
- ❖1870-90 — This period was humid in some areas and former Lake Ngami filled in the northwest of Botswana.
- ❖1875-1910 — There was a marked decrease in rainfall in southern Africa, and 1910 experienced a severe drought.
- ❖1921-30 — Severe droughts in the region.
- ❖1930-50 — Southern Africa experienced dry periods alternating with wet ones, and in some years the rains were very good. The 1946-47 season experienced a severe drought.
- ❖1950s — There was abnormally high rainfall in some parts of the region. East Africa experienced flooding, and Lake Victoria rose by several metres. Elsewhere, the equatorial region experienced below normal rainfall.
- ❖1967-73 — This six-year period was dry across the southern African region. The equatorial region experienced above average rainfall.
- ❖1974-80 — This period of six years was relatively moist over much of southern Africa. In 1974, the average annual rainfall was 100 percent above normal throughout the region.
- ❖1981-82 — Most of southern Africa experienced drought.
- ❖1982 — Most of sub-tropical Africa experienced drought.
- ❖1983 — This was a particularly bad drought year for the entire African continent.
- ❖1985 — Conditions improved.
- ❖1986-87 — Drought conditions returned.
- ❖1991-92 — Southern Africa, excluding Namibia, experienced the worst drought in living memory.

SOURCE: Tyson, P.D., *Climate Change and Variability in Southern Africa*, Oxford University Press, Cape Town, 1987

ried out in other countries in the region, with the same conclusions being drawn.

Regional weather systems

The Intertropical Convergence Zone (ITCZ) brings most of the rain that falls in the region. The ITCZ is a zone of intense rain-cloud development created when the Southeast Trade Winds (from the southern part of the region) collide with the Northeast Monsoons (winds from the north). The movement of the ITCZ southward away from the equator marks the start of the main rainy season in the Southern Hemisphere. This movement is linked to the position of the sun in relation to the region, which marks the seasons. During summer, when the sun is directly overhead between the equator and the Tropic of Capricorn, it heats the ocean and other water bodies. This causes warm, moist air to rise into the atmosphere, often resulting in substantial and rain-bearing clouds. During the summer months, the ITCZ is the main rain-bearing system over most of southern Africa.

One indicator of how well a season is performing is to monitor the position of the ITCZ, and compare this to its "normal" position at different phases during the rainy season. In a normal year, the ITCZ can fluctuate between mid-Tanzania and southern Zimbabwe, bringing good rains to most of southern Africa.

The ITCZ and other main rain-bearing systems have often been inactive in recent years and less effective in promoting rainfall. An atmospheric condition known as the Botswana Upper High also creates unfavourable conditions for heavy rainfall.[20] Its frequent occurrence almost always results in drought in some countries in the region. In some instances, like an expanding balloon, it tends to push the rain-bearing ITCZ and active westerly cloud-bands out of the region and over the Indian Ocean.

Rainfall variations across southern Africa

Rainfall in southern Africa comes almost entirely from evaporation over the Indian ocean. During winter and drought periods, the Botswana Upper High, along with the eastern mountain belt stretching from the Drakensberg in South Africa right up to Tanzania, blocks moist air from entering the region. The Botswana Upper High is a high pressure cell centred over Botswana (hence the name) between three and six kilometres above sea level. Its establishment is unfavourable to widespread rains across southern Africa.

Basic movements of air masses over southern Africa in summer Map 5.1

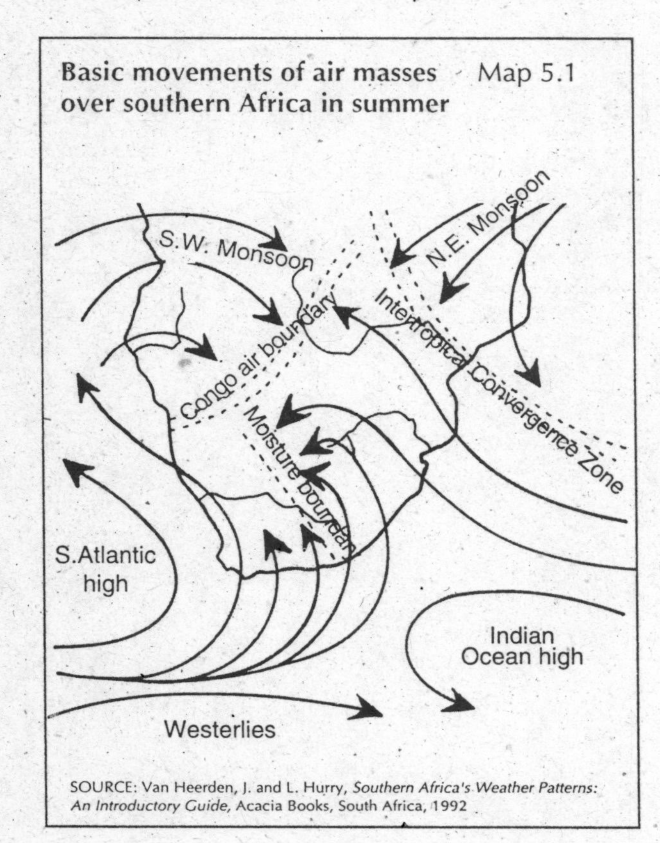

SOURCE: Van Heerden, J. and L. Hurry, *Southern Africa's Weather Patterns: An Introductory Guide*, Acacia Books, South Africa, 1992

Mean summer rainfall (mm) over southern Africa Map 5.2

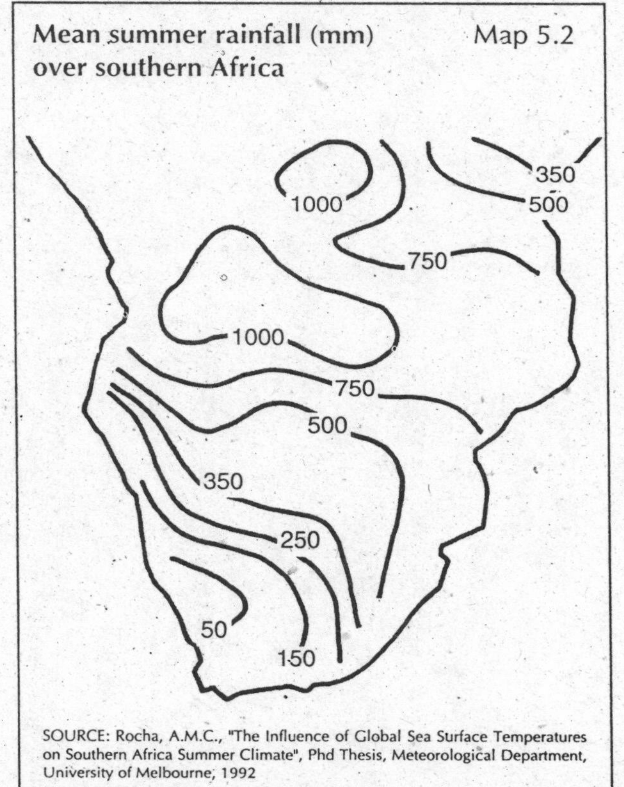

SOURCE: Rocha, A.M.C., "The Influence of Global Sea Surface Temperatures on Southern Africa Summer Climate", Phd Thesis, Meteorological Department, University of Melbourne, 1992

Basic movements of air masses over southern Africa in winter Map 5.3

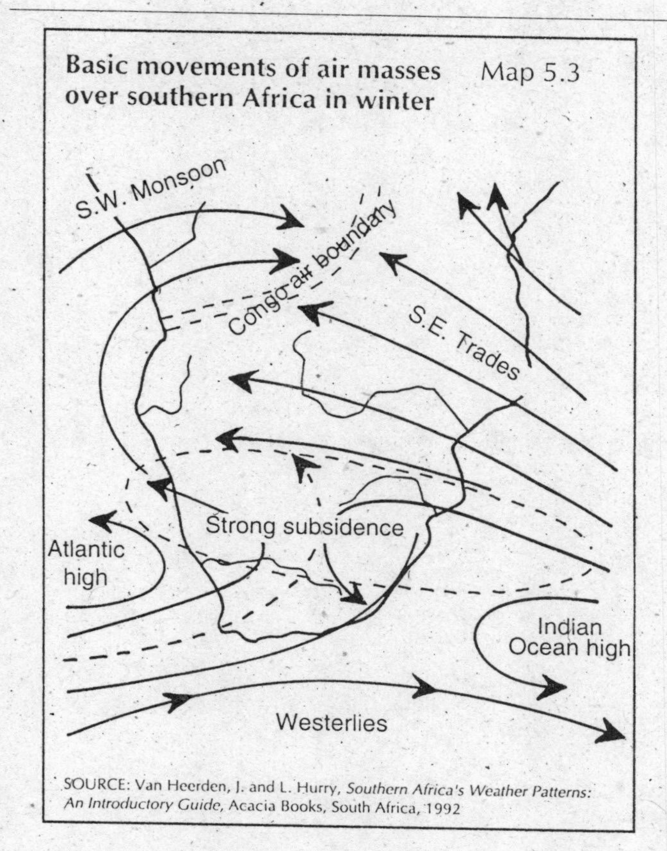

SOURCE: Van Heerden, J. and L. Hurry, *Southern Africa's Weather Patterns: An Introductory Guide*, Acacia Books, South Africa, 1992

Mean winter rainfall as a percentage of the annual total Map 5.4

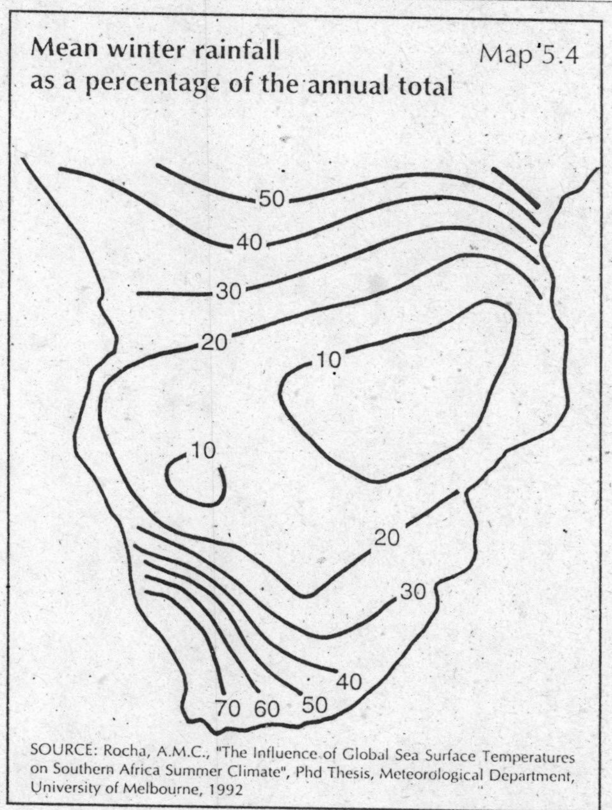

SOURCE: Rocha, A.M.C., "The Influence of Global Sea Surface Temperatures on Southern Africa Summer Climate", Phd Thesis, Meteorological Department, University of Melbourne, 1992

Moisture in the air generally increases from southwest to northeast in the region. Rainfall increases toward the equator, with the south and west of the region being arid or semi-arid.[21] The clouds gradually lose moisture as they move westward over the region, and less rain falls. This decrease in rainfall has an influence on rainfall variation, and areas with the least rainfall are most prone to high rainfall variability.[22] Examples are the nama-karoo, succulent karoo and desert. The dry savanna and fynbos zones have slightly more rain and less variation.

Malawi, located within the moist savanna zone, experiences fewer droughts because of the mountains which stimulate rainfall. The country's relatively good rains have made it a natural refuge for migrants from drought-prone areas.[23]

RAINFALL AND ECOZONES

Rainfall patterns, and the frequency and intensity of drought cycles, negatively affect the region's ecozones on a regular basis. The drier ecozones (the nama-karoo, succulent karoo, desert and large parts of the dry savanna) are particularly affected. Scientists studying these ecozones think that areas receiving less than 300-400 mm rainfall annually are controlled more by the short-term changes in rainfall than the long-term average. It is particularly so in areas where the amount of rainfall differs on a yearly basis from the average by more than one-third.[24] As the variation increases, so does the amount of change. The length of the drought seems to have a much stronger effect than its intensity. Plants and animals can survive a short, very dry spell more easily than a longer spell which is not as dry.[25]

Plants need water to survive and grow. Research has shown that in dry grasslands, the amount of grass cover can increase by up to seven times during a wet period.[26] In the desert ecozone, rainfall can vary by 100 percent from year to year, with the vegetation produced varying by up to 50 percent. During dry periods, some plants become extinct in their local areas while the same species thrive in areas of abundant rainfall. This may prevent the species from becoming extinct. The dry and wet periods also lead to changes in composition of plant species. Studies on dry grassland have shown variations of species composition directly related to dry and wet changes.

In areas which fall between the summer and winter rainfall areas, there seem to be opposing wet and dry cycles. When the winter rainfall area is in a dry cycle, the summer rainfall

Photo 5.3 PHOTOGRAPHIC TRAINING CENTRE, MAPUTO-R Rangel

People and wildlife adapt to rainfall patterns, changing their habits to suit the weather conditions.

Photo 5.4 SARDC-P Johnson

area is in a wet cycle. Parts of the nama-karoo area which fall between these two zones are affected by these cycles. Summer rainfall tends to favour grass, so during periods of high summer rainfall grasses dominate. But grasses go dormant in winter, so when winter rainfall is higher the shrubs are the winners.[27]

Animals which feed on plants are affected by these cycles. Their numbers increase and decrease dramatically in response to rainfall, which determines the amount of food available. Animals such as the wildebeest experience a population explosion during wet periods, followed by near-extinction during drought years — particularly where normal movements are restricted by fences and people.[28] This is also true for cattle, although artificial water-holes have helped many through a short drought.

The type of animals found in an ecozone also changes with the plant species composition. More grazers (grass-eaters) move into an area during a period of abundant grasses. Browsers (leaf-eaters) take over during times when leafy shrubs are dominant.[29] Large mammals tend to roam over large areas in search of food and water during drought. Elephants, which require large amounts of food, migrate in search of food and dig for water in dry river beds.

The vegetation in such regions has adapted to dry and variable environments. Some plants have specialised roots which store large amounts of water. Deep-rooting plants can obtain moisture at great depth and for longer periods. Examples of such plants are found in the Kgalagadi sands of Botswana, southern Zambia and western Zimbabwe where species such as the Zambezi teak remain in leaf throughout the year. Succulent plants, which have thick fleshy leaves or stems which store water, are also common in the region. Other plants are deciduous, shedding their leaves during the dry season to cut water loss through leaves. Still others can lie dormant for a long time until the next rains come, even a couple of years later.

FACTORS THAT WORSEN DROUGHT

Though droughts in southern Africa are regular climatic events, there is some question as to whether human activities may adversely affect weather patterns. Some scientists argue that severe deforestation or overgrazing over large areas may result in dry rather than fairly moist, warm air rising, enhancing the chances of drought. This view is contentious, as explained by a Canadian climatologist, who wrote in 1984 that while Africa's droughts are considered by most researchers to be aspects of natural fluctuation:

Weather monitoring and reporting in southern Africa Box 5.2

The World Meteorological Organisation (WMO) and the United Nations Food and Agricultural Organisation (FAO) assist southern Africa with weather monitoring.

The WMO project "Drought monitoring for eastern and southern Africa" consists of the drought-monitoring centres in Nairobi and Harare. FAO's involvement is mostly through the National Early Warning Units (NEWU) of SADC as well as through the Regional Early Warning Unit (REWU) based in Harare. Member countries send raw weather data to the centres. The centres in turn prepare bulletins such as the *Ten-Day Drought Watch for Southern Africa* and the monthly *Drought Monitoring Bulletin.*

Meteorological data is exchanged among member countries and the drought monitoring centres through WMO's Global Telecommunications system, by fax and telephone, but the performance of some modes of communication has not been satisfactory. Poor communication remains one of the major constraints to effective exchange of meteorological data in southern Africa.

Publications such as *Impact,* a magazine published by Climate Network Africa, inform on rainfall and distribution in southern Africa. Their information is supplied by the drought monitoring centres. REWU regularly publishes weather information of interest to farmers, policymakers, donor agencies and others concerned with food security.

While information and weather forecasting is available in the region, there has often been a lack of appreciation by government officials of the value of such information for planning purposes. Prior to the widespread drought of 1991-92, the drought monitoring centres advised all national weather services about the possibility of drought due to persistent anomalous atmospheric and oceanic behaviour, nine months before the eventual calamity. But some governments in the region did not act quickly, and were found wanting when the drought eventually came.

The challenge ahead is to ensure that weather information is translated into management decisions if crises are to be avoided.

SOURCE: Maurice Muchinda, "Global Weather Patterns in Relation to Southern Africa", Drought Monitoring Centre, Nairobi, Sept 1993

"... it is conceivable — though still unlikely in the view of some professionals — that human interference may be prolonging and intensifying the dry spells natural to the climate."[30]

There is no conclusive evidence linking poor land-use with decreased rainfall in southern Africa. Human activities do, however, increase the impacts of drought. There is agreement that the impacts are becoming more severe in southern Africa, and the environment more susceptible to irreversible damage as a result. Rapidly increasing numbers of people in southern Africa put more pressure on resources and lead to clearing of vegetation, use of marginal land, inappropriate land-uses and soil degradation, altering the landscape in the process. This has led to adverse environmental developments such as no groundwater recharge and flash floods.

Natural resource degradation

Droughts intensify land degradation and weaken the land's resilience — its ability to bounce back to full production after a drought episode. Overgrazing and poor cultivation practices in combination with drought can lead to the deterioration of pastures and arable land to the point where they must be abandoned.

There is little documented evidence yet to support the theory that overgrazing and expanding cultivation reduce the amount of rain that falls in the region. The more obvious result of overgrazing and overcultivation is soil erosion, floods, reduction of productivity and species shifts when the rains eventually return.[31]

Drought also directly results in the loss of trees and a lack of regeneration. Vegetation cover loss enhances the impact of drought in two major ways: the increase of rainfall runoff, and evaporation. Increased rainfall runoff affects many soil types, reducing nutrients, organic matter and moisture availability for the remaining plants. The increased daytime temperature range, because of the loss of vegetation cover, may also inhibit the regeneration or establishment of woody plants. In the end, woody plants using moisture from much deeper in the soil dominate at the expense of the more shallow-rooted grasses and trees which utilise moisture in the surface layers of the soil. All of this results in a significant change in species composition.

IMPACTS OF DROUGHT

Drought has its greatest impact on water supplies. Lack of water affects every aspect of environmental health and human activities, including agriculture, natural areas, industry and development projects.

Agriculture

Agriculture is the sector most directly affected by drought, and the impact can be extreme in dryland farming areas which rely on rain to provide water for crops. In Botswana, for example, where most cultivation is rain-fed, between 1979-1981 production fell from 60,000 tonnes to below 20,000 tonnes due to lack of rainfall. In 1984, production was as low as 7,000 tonnes. This represented a food decline from 715 to 100 kg per family as a result of drought.[32] In arable areas, problems of soil erosion are intensified and worsened during a drought. This is basically due to severe loss of vegetation during the dry spell.

Drought is the most important factor limiting livestock production in Africa. A single-season drought does not have a major impact on livestock numbers as the animals can survive, though they may not put on weight or breed. However, a drought lasting two or three seasons exhausts the available food sources, decimating herds. In Botswana, for example, drought reduced cattle numbers from 1.35 million to 900,000 in the dry years of 1964-67.[33] The 1991-92 drought, which ravaged most of southern Africa, killed more than one million cattle in Zimbabwe and many more in other coun-

Livestock and drought Box 5.3

Livestock can be severely affected during times of drought, and can die in large numbers under persistent dry conditions.

❖ During the 1980s drought, cattle in the Mopane area around Palapye and Francistown, Botswana, were in much better physical condition than those further south because *mopane* trees — which can be browsed — still had leaves.

❖ During the 1991-92 drought, Zimbabwe's commercial and communal farmers lost more than 1.5 million head of cattle, and agricultural experts believe it will be several years before the national herd fully recovers.

❖ In Lesotho, all livestock importation from South Africa is halted during a drought to give way to local sales of beasts. This is aimed at reducing grazing pressure on the rangelands.

❖ In Botswana, where cattle production constitutes a major economic base, a decrease in cattle population is a great loss. Rural people depend on their cattle for food and draught power. Loss of cattle forces them to hire tractors, which most cannot afford. In 1988-89, when the rains came, most communal farmers in Botswana had no draught power because their herds had been destroyed by the drought.

❖ Livestock numbers in the Kgalagadi have been flexible, adjusting to rainfall conditions. In the 1950s and 1960s, livestock increased rapidly. When drought set in, more than half of the livestock died. The 1970s brought another improvement in rainfall and an increase in livestock.

SOURCE: Zumer-Linder, M., "Botswana: Can Desert Encroachment be Stopped?", *Ecology Bulletin*, Vol 24, 1976, p. 179

Photo 5.5 SOUTHLIGHT-P Grendon

Intensive livestock farming can threaten fragile soils, especially in times of drought. Near Rietfontein in Northern Cape, 1990.

tries of the region. During a drought, overgrazing leads to further degradation of pastures and arable areas in cattle farming-areas. The deterioration of grazing capacity further reduces livestock numbers.

In drier areas, scanty rainfall for a few years can kill vegetation permanently and poor land-practices only make it worse.[34] In Namibia, for example, degradation of pasture is caused by interaction of overstocking and drought cycles.[35] Drought can also exacerbate deforestation, as some of those whose crops are lost attempt to make money by selling firewood.

Urban areas and industry

Drought also affects urban areas and industry. During the mid-1980s drought, the construction industry in Botswana was forced to reduce its activities after water reservoirs fell to critical levels. Beverage companies, which use a lot of water to wash bottles, had to change to non-returnable aluminium cans which require less water. The 1991-92 drought was no different. Botswana's construction and textile industries had to retrench workers after operations were scaled down because of a severe shortage of water. According to the Botswana Textile Manufacturers Association, 50 percent of the workforce in the sector was laid off during the drought.

Similar problems hit Bulawayo, the heart of Zimbabwe's industrial sector. Companies were almost forced to pull out and relocate elsewhere because of a lack of water, and half of the small businesses crumbled. In South Africa, Swaziland and Zimbabwe, sugar cane industries almost ground to a halt because there was no water for irrigation.

Rivers, lakes and groundwater

The drop in water supplies in dams and rivers also affects the quality of the water. Because of the reduced water volume, the concentration of sewage and other effluent in rivers increases, resulting in outbreaks of diseases such as diarrhoea, dysentery and cholera. The cholera outbreak that affected almost every country in the region during 1992 and 1993, claiming hundreds of lives, may have been compounded by the drought.

In many drought-affected areas in Zambia, streams and rivers dried up. Villagers, mainly women, had to walk long dis-

Photo 5.6 PHOTOGRAPHIC TRAINING CENTRE, MAPUTO- R Rangel

The life-sustaining Limpopo river, almost dry in time of drought in southern Mozambique.

Indigenous responses to drought: food reserves

Box 5.4

Over hundreds of years, people in southern Africa have developed effective responses to alleviate the ravages of drought on their communities.

The Basarwa (or San people) of northern Botswana share food as a form of "risk insurance" in times of drought. The relatively well-to-do share with the less fortunate, thus ensuring that everyone in the community survives. The Basarwa also increase their hunting and foraging activities to obtain extra food reserves. They have adapted to their harsh environment, surviving on very limited resources. They can live in areas receiving an annual rainfall below 350 mm, a minimum amount considered essential for dryland agriculture. The Basarwa can go for long periods without surface water, relying mostly on food and dew that collects on wild plants.

Rural people in Botswana utilise over 250 species of wild plants and animals for food in times of drought. Most of these plants are not used during years of good rainfall, and this is vital because it gives the plants time to recover and regenerate before the next drought. The Batswana have also developed food preservation and storage methods to establish and maintain a reliable, nutritious food-base during long-running droughts. The methods include sun-drying, salting, parching and fermenting foodstuffs.

Elsewhere in the region, subsistence farmers are also turning to drought-resistant crops. Tanzanian and Zambian farmers have adopted cassava, sorghum and millet as drought reserves. During the 1991-92 drought, the Government of Zimbabwe called upon farmers to grow drought-resistant crops, most of which had been abandoned for the more commercially viable maize.

Many rural people throughout the region ate wild roots and fruits to survive. These were plants that indigenous people had known for many generations as an alternative source of food during times of hardship.

At Sagambe village, Zimbabwe, more than 1,000 families survived the drought by supplementing their diet with wild plants. The villagers, who on occasion, supplement their staple diet of *sadza* (maize porridge) with pounded flour from a bush palm known as *nyamutata*, had to rely heavily on this brownish substance throughout the drought period. Many of the villagers crossed into neighbouring Mozambique to collect this bush palm from the Tangwena mountains.

Desperate villagers in the Southern Province of Zambia also resorted to similar actions to survive the drought. The villagers collected roots, some of which were poisonous, and either boiled or soaked them in water to get rid of the poison before eating them.

During an earlier drought in 1982-84, many rural households in southern Zimbabwe generated income by harvesting, shelling and selling wild *marula* nuts, a species found in several countries in the region. The villagers used the income to buy basic foodstuffs.

SOURCES: Fluret, Anne, "Indigenous Responses to Drought in Sub-Saharan Africa", *Disasters*, Vol 10, no 3, 1986
Muir-Leresche, Kay, "Drought Relief Strategies: Some Ideas", University of Zimbabwe, Harare, n.d., p. 4

tances, only to settle for polluted water. In many cases, the villagers were forced to share the water with wild animals and their livestock, making it even more unsuitable for human consumption. This is thought to have led to widespread outbreaks of dysentery and cholera that killed hundreds of people in Zambia in 1992. It is worth noting that when a previous cholera outbreak occurred in several countries in southern Africa in the mid-1980s, the region had just come out of another drought.

Drought problems go deeper than just the availability of surface water. The water-table is usually lowered if an area is hit by recurrent droughts. The lowering of the Kuiseb water-table, in the Namib desert, resulted in the wilting and

Rural water management conquers drought Story

Gaborone (SARDC) — Peter Mosweu has no formal training but the skill he has acquired over eight decades qualifies him as a rural water-management expert and drought survivor.

The 84-year-old farmer from Botswana, who never went to college or university, today hosts groups of students from a Gaborone agricultural college at his 51-hectare farm in Dinkgwana communal lands, to show them his water-harvesting skills.

"I did not go to school to train as a water engineer," says the grey-haired, five-foot Mosweu, "but I have equally good knowledge on how to harness water and I have used those skills to build two dams."

Fed up with the 10km trek his family had to make almost every other day to and from a district borehole in Mochudi village about 40 km north of the capital, Gaborone, the farmer built his first dam using a spade. The dam, on a fast-flowing underground stream, took three years to complete. It now holds about 400 cubic metres of water, according to local estimates.

Since the dam was built more than 15 years ago, Mosweu, his family and livestock have escaped largely unscathed by the cycles of drought that grip Botswana almost on a regular basis. During the 1991-92 drought, described as the most severe in living memory, Mosweu's family and about 100 neighbours survived with their animals intact.

While many people in Botswana lost some of their animals or could not use them as draught power because they were too weak, Mosweu's livestock — cattle, donkey, pigs, goats and sheep — was in good shape.

He supplies his neighbours with water "because it's good neighbourliness to do so." His philosophy is simple: Love thy neighbour. He explains that while he may have abundant water supplies today, he may suffer shortages in other necessities tomorrow and believes his neighbours will come to his aid.

Mosweu's fame as a rural water expert has attracted the attention of the Forum on Sustainable Agriculture, a Gaborone-based non-governmental organisation. It has since invited him to participate in its farmer-to-farmer programme whose aim is to get Batswana farmers to learn from each other's experiences and share skills.

Richard Kashweeka, the Forum's co-ordinator, said Mosweu's water-harvesting skills are of great benefit to other rural farmers. "Even in dry, dry seasons he has managed to feed his family, livestock and neighbours," said Kashweeka.

— **Munyaradzi Chenje,** Southern African Research and Documentation Centre (SARDC), Jan 1994

Photo 5.7

SOUTHLIGHT-P Weinberg

Traditional crops such as sorghum and millet are more drought-resistant than maize, which requires predictable rains or irrigation.

dying of acacia trees — a valuable and vital component of the lower Kuiseb ecosystem.

Human population movements

Historically, droughts have provided an impetus for mass migrations of people in southern Africa. People tend to move from an adversely affected area to a relatively better one. In Botswana, where boreholes dried up during the 1991-92 season, people moved first from rangelands to arable areas — and eventually to urban areas. Drought has also been a factor in the movement of commercial farmers off the land. South African commercial farmers numbered 120,000 in 1982 but only 66,000 remain on the land today. The rest have abandoned their farms due to drought — moving to urban areas and creating pressure on services and employment.

Wildlife and fish populations

The migration patterns of wild animals, including birds and mammals, are determined by seasonal rainfall. In the event of a drought, migrations are disrupted and wildlife numbers decrease, particularly herbivores. Severe loss of wildlife leads to ecological imbalances and economic losses.

Fluctuations in wildlife populations are fairly common in drought-prone areas. During a wet period, numbers increase rapidly, peak and then decline when drought sets in. The 1962 drought, for example, killed nearly all the zebra in Botswana. Hundreds of thousands of wildebeest also died.[36] The 1991-92 drought, which ravaged most of southern Africa, killed a significant number of wild animals in most countries of the region.

Fish populations also tend to decline during drought. Rivers and lakes shrink, and food sources for fish decrease, resulting in low breeding and smaller catches for fishermen. The characid (a fish found in the Zambezi), for example, breeds well in a three-to-four-month period when the river flows fast. During drought, the river flows slowly and there is little breeding.[37] Inland rivers and deltas such as the Okavango Delta in Botswana are a source of food and income for villagers in surrounding areas so the loss of fish increases the hardship of the drought.

PLANNING FOR DROUGHT

Climatic change and drought are recurring cycles in most parts of southern Africa. Drought has been studied here for over 100 years, and is recorded in text and oral history dating back many generations. It is reasonable, therefore, to expect and plan for droughts in this region — with or without global warming. Mismanagement of one drought leads to reduced productivity and greater susceptibility to the next drought.

Awareness, education and training can help to reduce drought susceptibility in the region. Today droughts are viewed as abnormal or unusual, particularly by urban populations, and not something for which people can plan or prepare. Droughts are also viewed entirely as environmental problems rather than problems exacerbated by people and their interaction with the environment. An improved knowledge base and increased understanding of all the factors surrounding drought is needed to improve awareness, education and training.

Early warning capabilities have been upgraded in the region, and information systems improved. Collection and dissemination of information can play an increasingly important role in drought anticipation and preparation. The challenge now is to encourage wider usage of this information for planning purposes.

Normal rainfall expected this year [1993]

Story

Johannesburg (ZIANA-Reuter) — Normal rainfall levels are expected over southern Africa this year after two seasons of severe drought that hit agriculture in the region very hard, weather experts said yesterday.

"Just about all the signs we look at are close to normal at the moment," weather researcher Simon Mason said in an interview. "The implications for the next two-to-three months are that the prospects for close to normal rains are very good and the chance of a bad start to the season or a particularly wet start are rather low."

The rainy season runs from October to April, covering an area running from around 10 degrees of latitude — from northern Angola, Zaire, Malawi and southern Tanzania through Zambia, Zimbabwe and Mozambique to the southern tip of the continent.

The region's weather is primarily governed by a massive area of equatorial ocean stretching from the Peruvian coast 10,000 km westwards to the international dateline in the central Pacific.

Surface warmth in the area, a phenomenon known as *El Nino,* spells drought for southern Africa. Conversely, an unusually cold surface current, *La Nina,* heralds unusually heavy rain.

Mason said a particularly strong *El Nino* developed at the start of the 1991-92 rainy season, bringing the worst drought of the century to much of southern Africa. Crops withered and hard-pressed governments were forced to spend scarce foreign currency to import food for their people.

South Africa, the main import point for drought relief for much of the region, has just received a soft loan of $850 million from the International Monetary Fund to help pay for emergency grain purchases.

El Nino conditions weakened in the first part of 1992, but suddenly and unexpectedly strengthened again last December, bringing a second successive year of drought, although less severe, in the 1992-93 rainy season to last April.

"This persisted through the first half of 1993, but there are indications that it weakened in July this year," said Michael Edwards, deputy director of the South African Weather Bureau. "Statistically over the past 70 years there has never been a period of three consecutive drought years — less than 75 percent of average rainfall over the main rainfall regions of South Africa."

He said there had been only two instances of two consecutive drought years during this period — between 1925-1927 and 1991-1993.

Unusually cold water in the western equatorial Indian Ocean means good rains, while warm water indicates drought conditions. Mason said the ocean temperature was now about normal after two years of fairly warm water.

—The Herald, Zimbabwe Newspapers, Harare, 29 Sep 1993

NOTES
CHAPTER 5

[1] UNEP, *The El Nino Phenomenon*, UNEP, Nairobi, 1992, p. 30-31
[2] Marsh, Alan and Mary K. Seely (eds.), *Oshanas, Sustaining People, Environment and Development in Central Owambo*, Government of Namibia, Windhoek, Jul 1992
[3] Whilhite, D. and M. Glantz, "Understanding the Drought Phenomenon: The Role of Definitions", *Water International*, 1985, Vol 10, p. 111
[4] Garanganga, Bradwell, "Drought Over Southern Africa: an Overview", for SARDC, Harare, 1993
[5] Falkenmark, Malin and Carl Widstrand, "Population and Water Resources: A Delicate Balance", *Population Bulletin*, Nov 1992, Vol 47, no 3, p. 18
[6] UNEP, op. cit. 1
[7] Garanganga, op. cit. 4
[8] Masika, R.S., "Rainfall Patterns in Eastern Africa", *Drought Network News*, Oct 1992, Vol 4, no 3, p. 7
[9] Tyson, P.D., *Climatic Change and Variability in Southern Africa*, Oxford University Press, Cape Town, 1987
[10] "The Drought", *Social Change and Development*, 1992, Vol 30, p. 18
[11] Tyson, op. cit. 9, p. 11
[12] Campbell, Alec, "The Use of Wild Food Plants, and Drought in Botswana", *Journal of Arid Environments*, Jul 1986, Vol 11, p. 81
[13] Tyson, op. cit. 9, p. 70, 197
[14] Huntley, Brian J. and others, *South African Environments into the 21st Century*, Tafelberg, Cape Town, 1990, p. 47
[15] Rocklifte-King, Geoffrey, *Drought, Agriculture and Rural Development: Policy Options*, Government of Botswana, Gaborone, Apr 1990, p. 3
[16] Boesen, Jannik (ed.) and others, *Tanzania: Crisis and Struggle for Survival*, SIAS, Uppsala, 1986, p. 146
[17] Teri, J.M. and A.Z. Mattee, *Drought Hazard in Tanzania*, Sokoine University of Agriculture, Arusha, May 1988, p. 115
[18] Tyson, op. cit. 9, p. 197
[19] Lema, Anderson J., *East African Climate: 1880-1980*, University of Dar es Salaam, Dar es Salaam, 1990, p. 271

Ibid.—same as previous note; op. cit. 5—same as note 5

[20] Matarira, C.H., "Drought Over Zimbabwe in a Regional and Global Context", *International Journal of Climatic Change*, 1990, Vol 10, p. 609-625
[21] Chabwela, H., *Wetlands: A Conservation Programme for Southern Africa*, IUCN/SADCC, Harare, Nov 1991, Vol 1, p. 11
[22] Boesen, op. cit. 16, p. 107
[23] Webster, J.B., "Drought and Migration: The Lake Malawi Littoral as a Region of Refuge", In *Symposium on Drought in Botswana*, Botswana Society, Gaborone, 1978, p. 155
[24] Ellis, James and others, "Climate Variability, Ecosystem Stability, and the Implications for Range and Livestock Development", in Behnke, Roy and others (eds.), *Range Ecology at Disequilibrium: New Models of Natural Variability and Pastoral Adaptation*, ODI, Nottingham, 1993, p. 33
[25] Behnke, Roy and others (eds.), *Range Ecology at Disequilibrium: New Models of Natural Variability and Pastoral Adaptation*, ODI, Nottingham, 1993, p. 36
[26] MacDonald, Ian and Robert Crawford (eds.), *Long Term Data Series Relating to Southern Africa's Renewable Natural Resources*, CSIR, Pretoria, 1988, p. 281
[27] Ibid., p. 286
[28] Campbell, Alec C., "The 1960's Drought in Botswana", In *Symposium on Drought in Botswana*, Botswana Society, Gaborone, 1978, p. 101
[29] MacDonald, op. cit. 26, p. 287
[30] Hare, Kenneth "Recent Climatic Experiences in Arid and Semi-arid Lands", *Desertification Control Bulletin*, May 1984, no 10, p. 21
[31] Muchinda, Maurice R., *The Variability and Reliability of Annual Rainfall in Zambia*, Meteorological Department, Lusaka, Zambia, Feb 1988, p. 1
[32] Morgan, Richard, "The Development and Application of a Drought Early Warning System in Botswana", *Disasters*, 1985, Vol 19, no 1, p. 44
[33] Tyson, op. cit. 9, p. 87.
[34] Muchinda, op. cit. 31
[35] Green, Reginald H., *Ecology, Poverty and Sustainability: Environmental Portents and Prospects in Rural Namibia*, Association of Agricultural Economists, Windhoek, 1990, p. 14
[36] Campbell, op. cit. 28
[37] Marshall, Brian E., "The Impacts of Introduced Sardine *Limnothrissa miodon* on the Ecology of Lake Kariba", *Biological Conservation*, 1991, Vol 55, p. 153

Section 2:

Risk and Capacity Assessment: Community-based Considerations

Section 2 comprises eight practical learning activities that aim at developing participants' understanding of different approaches to risk assessment in at-risk communities. Participants experience and practice a number of different processes for gathering and analysing data, including visual mapping, participatory rural appraisal and questionnaire administration. There are a host of techniques for assessing risk. These activities focus on those most suitable for use at community level. They aim to improve participants' learning skills, particularly those which assess risks and capacities at individual, household and community levels.

How do we construct hazard, risk and capacity maps?

Purpose

This activity assists participants in generating information about specific hazards, vulnerabilities and capacities in at-risk Southern African communities.

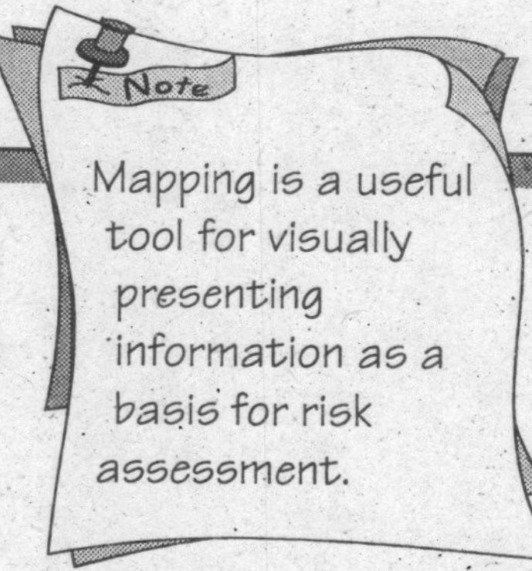

Note
Mapping is a useful tool for visually presenting information as a basis for risk assessment.

Procedure

This activity asks participants to draw maps of familiar areas and communities, and to identify known risks and capacities.

Time:

- 1½ - 2 hours

Process

Introduction: Area Mapping

1. Introduce the activity by outlining the purpose and procedure.
2. Ask participants to work in 'area groups', i.e. people from the same country/ geographic area should work together.
3. Hand out big sheets of flipchart paper and coloured pens and ask groups to establish themselves around a surface such as a table, or floor area.

Participant Action

1. Give the following instruction:

   Draw a map of a particular area that you are familiar with within your country. The map should be as detailed as possible. Indicate the following on the maps:
   - *hazards*
   - *vulnerabilities*
   - *capacities and resources.*

   (You have 40 minutes to complete this task.)

2. Suggest that participants may want to use symbols to indicate hazards, vulnerabilities and capacities. Remind them to develop a key (or legend) which explains the symbols.
3. Move around the groups to monitor progress. Check that participants do not only focus on physical features of their area. Emphasise the importance of including human elements.

7. Ask groups to display their maps on the walls.

8. Request that a representative from each group presents and explains the maps to other participants.

Review and Discussion

1. In plenary, discuss common features and differences as revealed on the maps.

2. List identified hazards on flipchart and mark those that are typical for your area and the Southern African region as a whole.

3. Discuss how vulnerabilities and capacities relate to given hazards and identify elements most at risk.

4. Focus on the sources of information participants drew on in order to develop their maps. Ask questions such as the following:
   - ? Where did you get the data for your maps from?
   - ? How do you know your information is accurate?
   - ? What additional information would you find useful, and where would you find such information?
   - ? How are the different types of information linked to different sources?
   - ? Which sources of information are particularly useful for hazard, vulnerability and capacity assessment? Why? How?

5. Ask participants whether they have gained new insights as a result of the presentations, questions and discussion. Point out how discussions generate information and how people are a crucial source of useful data.

6. Encourage participants to apply their new information and insights to their maps by adding additional data. (Allow approx. 10 minutes for the review of maps)

Introduction: Community Mapping

1. Point out that the area maps show the general pattern of risks and capacities; the information thus gathered is useful for planning policy or large-scale initiatives. However, in order to target the elements most at risk, more specific and detailed information is necessary. The next activity will therefore ask participants to hone in on their area, and focus on a specific community.

Participant Action

1. Give the following instructions:

> — participants will work in their area groups and, together, draw a new map;
>
> — one person will act as the informant: s/he will draw on her/his knowledge of a familiar community in order to give information in response to questions;
>
> — the other person(s) will question her/him;
>
> (You have 20 minutes for this task.)

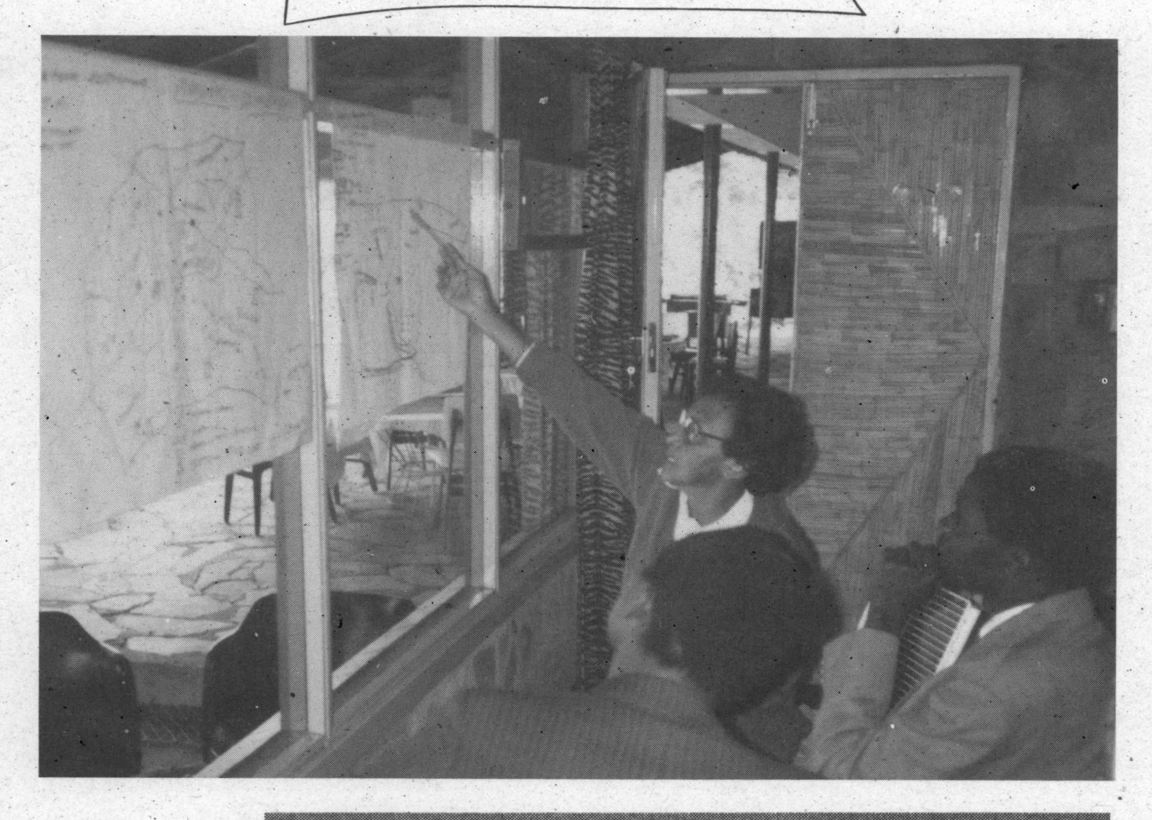

2. Monitor the progress and assist where necessary.

3. Ask groups to display their maps on the walls.

4. Request that a representative from each group presents and explains their community maps to the other participants.

Review and Discussion

1. Compare the area and community maps with each other: discuss common features and differences.

2. Ask participants whether the hazards identified on the area maps affect all communities the same?

   ? What factors would participants need to consider in order to assess the potential impact of the hazard on various communities?

   ? What were the different indicators of vulnerability on the area and community maps?

3. Ask participants how the community maps improved their understanding of the risks faced by communities.

4. Point out that the next stage of assessments would have to be at the *household level*. This would generate information about the vulnerabilities of specific groups of people, such as women and children.

5. Ask participants to list the following:
   What type of things would you look for in order to conduct a risk and capacity assessment at the *household level*?

   Answers may include the following:
   access to water
   access to livestock
   access to land
   source of fuel
   family remittances
   health status
   dependent members of households

6. Point out that this information would be very specific and could not be generated in a classroom situation, as the outcome would be too generalised. Refer participants to 'PRA' processes, such as 'wealth ranking'.

7. Facilitate a review of the activity. Explore what participants learnt from this activity.
   - ? What did they find surprising?
   - ? What was new?
   - ? In what way did the activity improve their understanding of their areas and the region?
   - ? How did it improve their understanding of the risks faced by vulnerable communities?

8. Summarise by drawing general conclusions about the hazard and vulnerability profile of their areas and the region.

Discuss what other kinds of maps you could develop: e.g. maps specifically focusing on the risks of women and children at the individual, household and community levels.

Learners' Responses

Review and Evaluation

Participants responded very positively to this activity. Their comments may guide your line of questioning when reviewing the process:

- Surprise at our own capacity to produce knowledge and information that is useful for assessing and planning: we don't have to rely on 'experts' to assess risks and vulnerabilities.
- No external expert can plan in a new environment: "You have to live in a kraal to really know the resources and vulnerabilities."
- No development work can be planned in isolation from the sources, and people who live in an area have knowledge of underlying factors which must be considered.
- Visualising elements brings them out into the open.
- Planning should be built on existing resources: itemising (in drawings) what there is, gives us a picture of elements we might have overlooked.
- There are different kinds of information, and some at the household and community level is crucial, but you can only access it if you ask and work with the people at that level.
- Mapping is a useful way of learning about a new area.

How can community-based research lead to risk assessment?[1]

Purpose

This session is a basic introduction to the what, why and how of Participatory Rural Appraisal procedures.

This introduction should ideally be presented by someone who has experience with PRA in the field. The session does not aim at training but information-giving; all it can attempt to do is give a descriptive overview of the approach as an alternative, truly community-based research and planning process, and thereby whet participants' appetites for participating in training in the field.

Procedure

Much of the session is a question-and-answer information giving process; facilitators may decide to include brief exercises based on a 'timeline', or a venn diagram, using participants as informants.

Time

- 2 hours

Materials

- visual illustrations (sample: see resources)
- information on PRA processes (see resources)
- reading: extract from "PRA report in Tete" (see resources)

Process

Introduction

1. Introduce the session by outlining the purpose and procedure.

2. Begin with story-telling: outline a case scenario of a community in need of aid and development assistance.

Sample Story

In August 1994, several hundred refugees returned from neighbouring Zambia and Malawi to Ndzadzo, Tete province, Mozambique. They had been away for 2-3 years and they brought little with them: some cooking utensils and tools - and the lucky ones a bicycle or meagre supplies of grain. They returned to a countryside where nature had taken over, erradicating any signs of previous cultivation. They settled next to the road, on the land allocated, and began to rebuild their lives. All the villagers were malnourished and many sick; life was one ofextreme poverty, and it was clear that these returning refugees needed assistance - but what was needed most urgently?

3. Ask participants how they would go about assessing key vulnerabilities and capacities related to the risks faced by a community such as the one in the sample.

4. Point out that conventional processes such as surveys and questionnaire-based interviews, administered by international or regional aid agencies, would have one thing in common: the information generated would reflect the bias of the surveyers rather than the perspective of affected villagers.

   Explain that Participatory Rural Appraisal (PRA) is a process in which community members themselves gather and analyse the information necessary to make planning decisions. Point out that this information would reflect what is meaningful and useful for community members: in determining risk, *their perceptions, not ours*, determine priorities.

Example

A *survey of the living conditions of a poor community without sanitation might suggest that the community may face an increased risk of diarrheoa. A risk reduction plan could therefore target sanitation and health education.*

However, a PRA revealed the following perceptions and priorities of risk: while the villagers of Ndzadzo were acutely aware of the need for sanitation they pointed out that at this stage it was not a priority as there was no overcrowding. Instead their *most urgent need was food and water. They explained how insufficient supplies and poor quality of water and food caused a lot of diarrhoea. Worse: childen suffered from chronic chest infections because of inadequate shelter. In particular those returnees who had arrived last were at risk, as they had not been able to mud the walls of their houses due to lack of water.*

It became evident that the most urgent demand was for water - rather than toilets or health education.

5. Describe some of the specific features of community-based assessments such as participatory rural appraisal:
   - PRA begins with the question 'Whose knowledge counts?'; the process acknowledges that poor people are capable and creative and need to be active partners in development;
   - PRA is a response to the dissatisfaction with aid and development interventions that are based on the perspective of outsiders rather than vulnerable communities themselves, and that often create dependencies rather than developing existing capacities;
   - PRA was developed in the South, in the field, amongst at-risk communities, and it was spread South — South;

- PRA necessitates a shift in attitudes and behaviour towards listening, standing back, being respectful and requesting to be taught, and open to learning;
- PRA activities fall into two groups: those primarily aimed at producing information, and those primarily aimed at assessing information through comparisons, ranking, etc;
- PRA aims at action: data are generated and analysed for understanding, and used as the basis for development planning.

6. Explain that a PRA can generate information that is sensitive to the specific risks and capacities of a community. If we aim at community-based disaster reduction it is crucial that we base collaboration on shared understanding with vulnerable communities.

   Give examples to illustrate your point.

Examples

** The time line revealed that the area is prone to the* ***hazard of drought,*** *and each of the seven previous droughts has it's own name describing the particular* ***coping strategy*** *used by the community to deal with the drought*

** The community map illustrated how water sources were scarce thus contributing to the* ***vulnerability*** *of household members to disease.*

** The seasonality diagram showed the times of the year when different wild crops such as wild fruit, roots and insects are harvested;**(capacities)**..*

** The activity clocks allowed a comparison of men's and women's duties and lead to an analyisis of* ***gendered vulnerability.***

** Childen's drawings were a source of information about the mental health status: the severe trauma suffered as a result of civil violence and war and the ongoing* ***risk*** *due to landmines.*

They also provided a key to understanding children's hopes and dreams for an improved life.

Participant Action

1. Ask participants to get into three discussion groups, and distribute copies of the seasonality diagram, children's drawing and community map.

2. Give the following instruction:

> *As a group, choose one of the data sheets.*
>
> *1 Describe the data presented:*
>
> *What do you learn about the community?*
>
> *2 Identify the risks and capacities:*
>
> *What are the vulnerabilities? What are resources?*
>
> *3 Analyse them in terms of priorities:*
>
> *Which risks/capacities seem most prominent? How/ why?*
>
> *You have 20 minutes to complete the task.*

3. Monitor the process and assist where necessary.

Review and Discussion

1. Facilitate report backs from the group activity; list important points on flipchart..
2. Briefly, outline the process that lead to the production of the data sheets. Describe how a seasonality diagram and a mapping exercise may be initiated and run, using available resources and being sensitive to considerations of literacy. (see resources)
3. Initiate a discussion on the following question:

   How can we use the information generated in order to plan risk reduction programmes?
4. Sum up the activity by giving an example in order to illustrate how key findings from the assessment process can lead to risk reduction plans.

Example: the case of Ndzadzo:

Older members of the community had a clear understanding of the importance of drought tolerant seed cultivation, and all villagers expressed their desire to actively cultivate their fields, rather than remain passive recipients of food aid. Based on these capacities a sustainable community development programme which would reduce drought related risks could include: distribution of drought tolerant seed such as sorghum in time for the planting season, linked to the establishment of a community seed-bank.

1. There is a wealth of reading and reference material on RRA / PRA / PALM procedures; a facilitator who has no previous experience with PRAs but wishes to conduct this session would be well advised to read numerous reports on PRA procedures conducted with vulnerable communities in order to be able to present an informed picture of how this approach can generate useful and representative information and project plans. A useful reference for reports and manuals is the International Institute for Environment and Development (IIED) 3 Endsleigh Street, London, WCIH ODD, UK.

Resources

VISUAL ILLUSTRATIONS

Seasonality Diagram

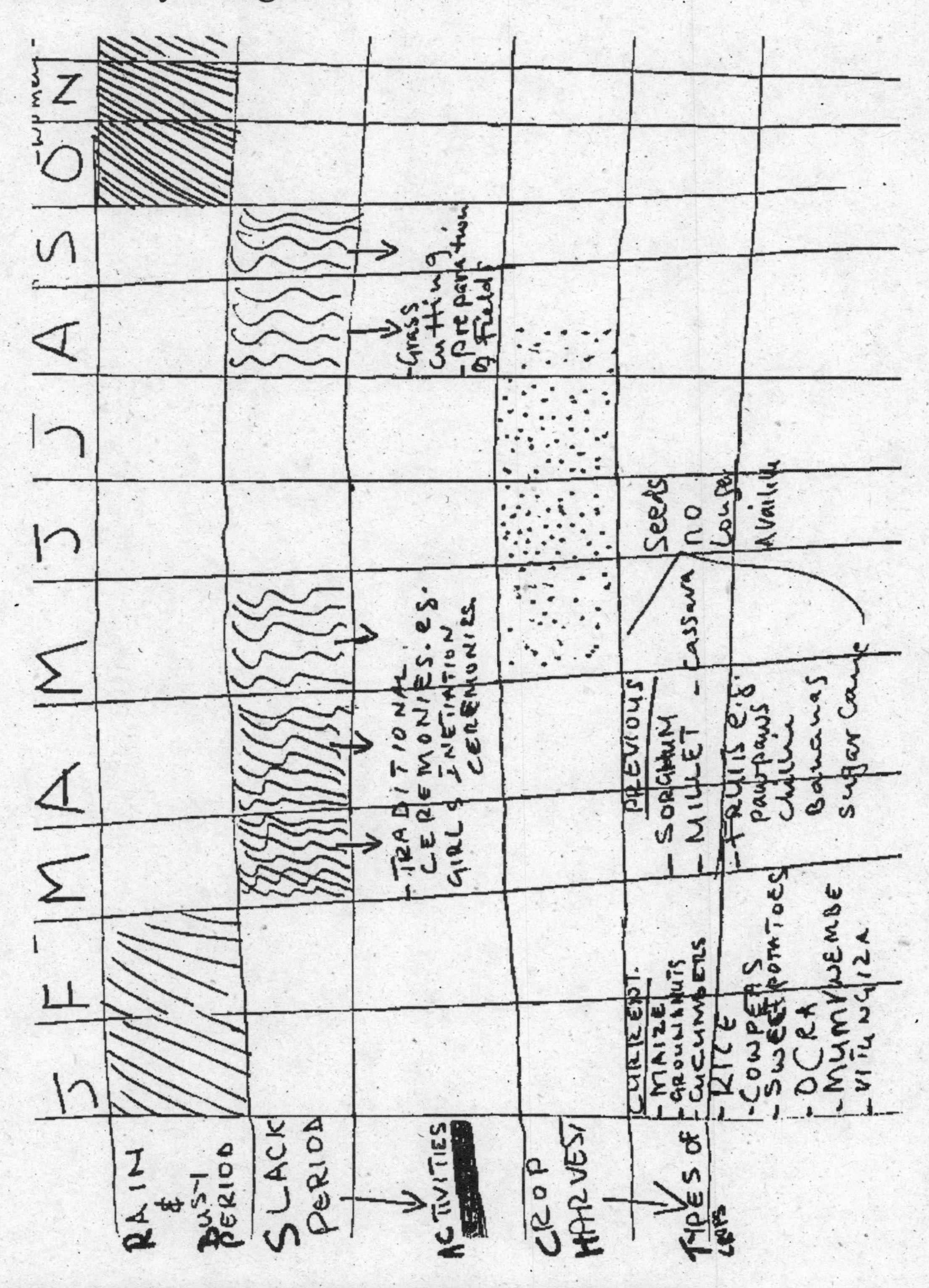

Children's Drawing

Community Map

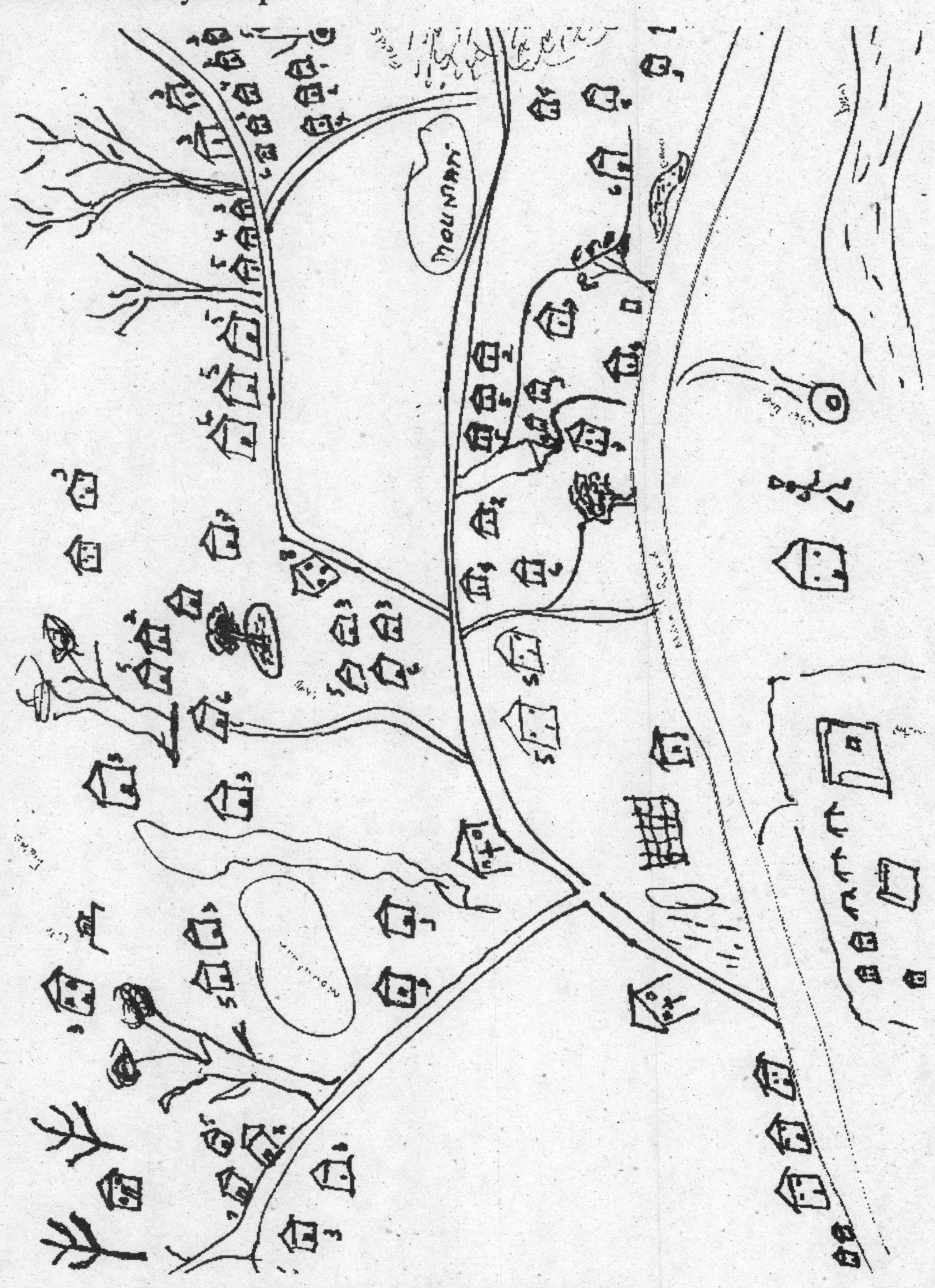

Information on PRA Processes

Mapping

Participatory mapping is simply asking community members to draw an area, showing key features and landmarks as perceived by the participants.

- Decide what kind of a map should be drawn (eg showing physical / geographic features, households where disabled people live, available community resources).
- Speak to community members: say you would like to learn more about the community, and particularly the aspect you have chosen. Ask them to identify a suitable area (eg. a large flat ground suitable for drawing outlines in the sand) for conducting the mapping exercise.
- Explain to everyone present what the purpose of this exercise is.
- Sit back, relax and observe what happens. Don't interrupt or interfere: be patient!
- Ensure that the map is copied as a permanent record (on paper); if applicable, record the names of the main participants who facilitated and managed the process.

Seasonal Diagramming

This is a calendar showing the main activities performed during the course of a year by different members of a community.

- Decide what kind of a diagram should be drawn (eg. rainfall, labour demand, diseases, crops harvested or wild fruits gathered).
- Speak to community members who have knowledge of the issue being investigated, and who would like to share their knowledge. Depending on the aspect chosen (eg. wild fruit gathering) you may wish to work only with a group of women.
- Explain what the purpose of the exercise is, and what you would like them to do, i.e. show how certain things change throughout the year.
- Draw up a 12 - 18 month calendar, but let participants choose where in the year they want to start.
- Encourage participants to use available materials - such as sticks or stones or leaves - to show when in the year there are more or less rainstorms / people working / crops harvested.
- If the exercise is done on the ground produce a paper copy as a permanent record.

Children's Drawings

Children do not need much instruction in order to begin drawing - they enjoy the process and they will spontaneously produce what is important to them.

- Children usually assemble near adults' gatherings: this may be a good time to initiate drawings. Speak to them: say you would like to learn more about their lives.
- Identify a suitable place, preferably with a reasonably hard surface.
- Provide sheets of flipchart and thick marking pens or crayons.
- Stand back, relax - don't interfere.
- If there are numerous occasions when children draw you may want to encourage equal participation by girls, and different age groups - this will lead to a greater variety in perspective.
- Review the drawing with the children; ask for explanations if necessary.

READING

Participatory Rural Appraisal (PRA) and How it Relates to Disaster Management[1]

Today we are acutely aware that community participation is key in the success of development programmes targeted at vulnerable communities. We have also come to understand that community health and environmental conservation strategies can only be sustainable if they are planned and implemented with the full participation of the people they are intended to benefit. Today, we would never question the argument which stresses that community involvement is a pre-condition for successful and lasting development.

But what about community involvement in disasters? Is a disaster not defined as an event which outstrips the capacity of a community to cope with it? What comes to mind?......... images of fire engines, food relief trucks, supplementary feeding and emergency medical teams rushing to the aid of the affected? .... communities overwhelmed? .... Let's just think about this for a while. Let's go back to the great Southern African drought of 1991/92, or the floods in Cape town in June 1994, and Cyclone Nadya which struck Mozambique's northern coastline in March the same year.

What about the hundreds of thousands of internally displaced and refugee Mozambicans who returned to villages destroyed after nearly two decades of hostilities? Were the communities affected by these events merely passive recipients of international and government aid? Have they become so dependent on outside help that they are completely disinterested in taking measures themselves which could help them be better prepared? Is their educational level considered to be so poor that such communities could never grasp the concepts of disaster prevention, mitigation and preparedness?

In this paper, we intend to show how Participatory Rural Appraisal can be one useful tool in assessing the hazards, vulnerabilities and capacities of disaster-prone communities based on a field experience with refugee returnees in Tete province, Mozambique. We hope to show that it is a valuable method for both empowering disaster prone communities to reduce their risk to known threats, and to improve programme planning by outside agencies, for instance, non-governmental organisations. Beforehand though, let's go over some basic terms commonly used. Disaster management is often presented - or interpreted in three rather different ways. First, some people view it mainly as disaster preparedness, or a readiness to respond in a timely way to a known threat. Often, this includes contingency planning as well as having essential relief supplies (i.e. tents, blankets, cooking sets, jerry cans, essential health kits, etc.) prepositioned for a rapid response.

Second, some equate it with emergency management or the actual response once a crisis has occurred, particularly the actions taken to minimise loss of life and property (most often seen in rapid onset disasters).

A third interpretation more commonly referred to today is disaster reduction. This is rather an embracing term for all measures which reduce the human, property, material and related losses caused by disasters. It gives particular emphasis to the importance of disaster prevention, mitigation and preparedness, and the need to incorporate these into ongoing development programmes and strategies.

There are other terms that are frequently used in disaster management:

Hazard:

This is an event or occurrence that has the potential for causing injury to life, or damage to property or the environment on which a community depends for its social and economic existence. Some examples of relevant hazards in Southern Africa include drought, civil unrest/conflict and epidemic dysentery.

Vulnerability:

This is usually viewed as a set of prevailing conditions or elements which adversely affects an individual, a household or community's ability to cope with a threatening event or process. For example, in Southern Africa, where many villagers live in semi-arid zones, drought is a recurrent threat. In some communities, agricultural extension workers provide support and training in drought tolerant seed cultivation. Others, which are isolated far from district service centres, may be excluded from both the technical support and the availability of drought tolerant seed. They are more vulnerable to the effects of drought.

Capacity:

These are the qualities which increase the ability of an individual or community to cope with a threatening event or process. Those drought prone communities who have an open attitude to cultivating small grains perhaps have a greater capacity to cope with repeated rainfall failure. Similarly, those communities who have the social/organisational capacity to run a revolving seed bank can also cope better.

Risk:

Understanding this term is critical to reducing the effects of natural and other threats. Theoretically, a risk = hazard + vulnerability + elements at risk. It is the anticipated losses (lives lost, numbers injured, property damage and disruption of economic activity) from the impact of a given hazard on a given element over a specific period of time.

Let's take flood-risk as an example. In an urban area prone to flooding, some houses have been constructed in a low-lying area close to the river bank. These are made of concrete block and have basements. Other houses made of corrugated iron, cardboard and thatch have been constructed in a dry river bed. When heavy rains fall up-stream, this hazard does not affect the houses or their occupants equally - although they are affected by the same rain fall rate. If flooding occurs, the water may wash through the basements of the concrete buildings, but leave the structures reasonably intact. But, in the river bed, the fragile dwellings are completely destroyed. It is the economic vulnerability of the riverbed dwellers that has forced them to live in this dangerous site. Moreover, their property is structurally more vulnerable than the concrete buildings of their neighbours. It is the vulnerability that has increased their risk, not the hazard, alone.

A key objective in disaster management, is to reduce risks. This can be done in many ways. For instance, in the flood example above, the risk might be reduced by civil engineering measures to control river flow rate up-stream during the rainy season. These steps can reduce the hazard. However, expanding employment opportunities for the riverbed dwellers, or relocating them to structurally sound accommodations outside the river bed would lower their vulnerability to a seasonal threat. Either strategy reduces the flood risk.

One important dimension to risk is understanding how people's perceptions of risk and their priorities vary. For instance, urban visitors to a rural community without sanitation might perceive that the community faces an increased risk of say, diarrhoeal diseases (due to lack of latrines or toilets). "Yes", the community might agree, "diarrhoea is a risk - but it is not our priority as there is more hunger here than diarrhoea". In determining risk, we must understand what the community perceives and prioritises as risks, from its perspective, not ours.

Participatory Rural Appraisal is a key assessment tool to help choose disaster reduction strategies that are meaningful and useful for vulnerable communities. It is one method which enables us to set shared priorities with a community, and choose strategies which build on existing community strengths.

To illustrate the role PRA can play in disaster reduction, we will refer to a PRA conducted in rural Mozambique in 1994. Three days were spent working with several communities, followed by a review of the information gathered. Listed below, is an example of how the methods we used generated information for programme planning that is risk sensitive (specifically, drought sensitive).

(Methods shown in *italics*, interpretation shown by underlining)

— From the *time line*, we learned the area is prone to the hazard of drought

— From the *time line*, we also learned that each drought has a specific name depicting a particular coping strategy (capacity) used by the community to deal with drought (the hazard).

— The *time trend* helped us to understand the rain patterns during a good harvest year, compared with rain patterns during a bad harvest year. This also helped us to understand the hazard's impact on crop production.

— From the *time trend*, we also gained an understanding of when the work in the field is at a high or at a low (key for planning community meetings or other gatherings that would take time away from cultivation).

- From the *seasonality mapping*, the community showed us the times of the year they collect wild foods such as fruits, nuts, roots etc. (capacity).
- The *transect walk* gave a better understanding of the map done by the community and an opportunity to ask questions. In addition, it provided further information in the community's vulnerabilities and capacities - such as a tinsmith making cooking pots.
- From the *community mapping*, we learned where the water sources were located, allowing us to better understand the drought related vulnerability faced by the villagers.
- The needs matrix provided an opportunity for us to understand the priority needs identified by the community - food, water and medicine.

Other examples of capacities and vulnerabilities were identified.

From PRA to Disaster Reduction Planning

With this information, we can already begin our programme planning - which ideally, should be carried out jointly with the community. From a disaster management perspective, some of the key findings that should be incorporated into the programme planning are listed below: -

- This area is drought prone.
- Hunger/food insecurity is perceived as the most important risk facing the community. -
- Lack of safe and accessible water as well as essential medicines are perceived as key vulnerabilities by the community.
- Older members of the community have a strong knowledge of and openness to using drought tolerant crops (capacities in attitude and skills).
- Older members of the community have a strong knowledge about the availability and use of wild foods (another capacity).
- The community members indicated which months they would be fully committed to preparing and cultivating their fields, highlighting another capacity to protect household food security, which any outside intervention should not undermine.
- Community members identified that limited access to essential medicines was a vulnerability, but could not associate this with any particular disease(s). This in fact reflects at least one other vulnerability as well as a capacity.

The second vulnerability is knowledge/skill based, in that they have not had access to health education which links disease to particular preventative or even simple curative interventions (at least biomedically, although there may be a host of traditional methods they are familiar with). At the same time, there is a strong expressed interest related to health matters a real capacity to explore, and build on, in the future.

From these findings, we could begin to develop a programme which would reduce the community's vulnerability, by reducing the impact of recurrent drought. For example, let's build on two of the four capacities identified above - the awareness among older members of the community of the importance of drought tolerant seed cultivation and secondly, the clear commitment by the community to actively cultivate their fields (not interested in remaining passive recipients of food aid).

A sustainable community development programme which would reduce the drought related vulnerability of the community might have the three following components:

Start up distribution of drought tolerant seed (e.g. sorghum) as well as maize and groundnut seed, linked to a community managed seed bank. This could also be supported by skills training of young people in local cultivation, involving the older members of the community, keeping in mind that most of the youth have spent their formative years in refugee camps, where food was delivered by relief agencies, and not cultivated by the household.

PRA as a Method for Planning Risk Reduction

PRA, as an approach to assess risks as well as vulnerabilities and capacities, is a valuable method in disaster reduction planning at community level. When applied in Tete Province in Mozambique, some of the tools used included time line, time trend, needs matrix, transect walks, community mapping and seasonality mapping. We will now show how the information collected through these methods helps an outsider gain a clearer understanding of drought-related vulnerability for this returnee population.

From the time line, events that had occurred in the community were tracked over many years. The hazard of drought was a recurrent event in the community's past. However from both the time line and the seasonality mapping exercises, it became clear that the community had coped with this known hazard in the past.

As an example, in 1973 during the Kansale drought, the community survived by eating a particular type of wild fruit, hence the name given to this drought. At this time, the community was also cultivating drought-tolerant crops such as sorghum. So, yes, the community was drought affected, but fully able to manage without external assistance.

Today, however, the situation has changed. The community is in the process of re-establishing itself after several years in refugee camps in neighbouring countries. As part of a repatriation/rehabilitation programme, a number of outside organisations have developed programmes to put these communities back on their feet. However, it appears that many of these programmes were carried out without a complete picture of the area's risk profile (especially the fact that it is drought-prone). Moreover, there appeared to be little understanding of the past capacities to deal with this threat, and how these needed to be supported/strengthened at least temporarily, as the returnee communities re-established themselves.

As a result, the rehabilitation programmes although well intended, have not been as effective as they might have been in reducing the vulnerability of these repatriating communities. On one hand, the returnees did receive seeds. However these were non drought tolerant maize seeds and were delivered late for planting. Therefore, although the villagers planted the maize, the crop failed because of prevailing drought conditions. Concurrently, the community has been supported with external food aid. However, this help is planned to end in accordance with a predetermined time frame set for returnee assistance, that never factored in the drought profile of this area.

For members of this community, drought as a hazard has not changed. But the risk had actually increased, as they are no longer able to cope with its consequences as they did in the past due to the considerably diminished capacities identified in the course of the PRA.

From a disaster reduction perspective, PRA is one tool which can be used to assess key vulnerabilities and capacities, as these relate to the risks faced by disaster prone communities. Compared to other assessment methods, PRA is particularly powerful, as it;

- actively involves the community
- empowers the community to identify the risks and priorities, as well as capacities to reduce these risks.
- provides a picture of the community's perceptions of the risks it faces
- allows both community insiders and outsiders to jointly identify risk reduction measures - is both time and cost effective

In this example, PRA provided a wealth of information, directly relevant to the risks affecting the returnee communities, and the capacities which could be strengthened to reduce these. This type of information is critical if outside agencies are to develop programmes which lower both disaster risk and community vulnerability for the long term. Southern Africa is a region with growing numbers of households at-risk due to socio-economic, environmental and climatic forces. However, over the past two years, PRA has become a practical tool which growing numbers of NGOs now use for project planning at community level. In fields as diverse as environmental conservation, home-based care for AIDS patients and community seed bank management, PRA is increasingly viewed as a method which enables NGOs and their clients to jointly plan programmes which make inroads into community vulnerability. The Southern Africa Regional Delegation of the International Federation of Red Cross and Red Crescent Societies is one of many humanitarian agencies which actively promotes the use of PRA for community assessment, and has advocated its use in community-based disaster mitigation planning.

Ailsa Holloway and Diane Lindsey, International Federation of Red Cross / Red Crescent Socity, Regional Delegation for Southern Africa, Harare.

How do we conduct vulnerability and capacity assessments?

Purpose

This activity will develop participants' ability to critically explore vulnerabilities within familiar communities.

The activity and ensuing discussion will lead to a deeper understanding of some of the problems with conducting vulnerability and capacity assessments at a distance from the communities concerned.

Procedure

Participants complete worksheets that lead to the construction of profiles of familiar communities.

Time

- 1½ hours

Materials

- pens and paper
- worksheets for each participant (see resources)

Process

Introduction

1. Outline the purpose and process of this activity.

Participant Action

1. Hand each participant a copy of the worksheet. Give the following instruction:

> *TASK 1:*
>
> *— Please complete the worksheet, working individually.*
>
> *— You will have 20 minutes for this task*
>
> *TASK 2:*
>
> *— Move into small groups (of 3-4)*
>
> *— In groups, give 5 minute presentations of your work*
>
> *— Answer questions of clarification.*
>
> *Prepare for a plenary discussion.*
>
> *You have 40 minutes for these tasks.*

2. Monitor the process of participants' individual progress and presentations in groups. Assist if necessary. Ensure that time limits are observed.

Review and Discussion

1. In plenary, initiate a review of both the process and the results of the vulnerability /capacity study. Ask participants, "What were the main problems you faced in managing the task?"

2. Point out the problems with conducting a vulnerability assessment away from the community it is based on, and without using a participatory process. Ask questions such as the following:

   ? What sources of information did you draw on in order to construct the profile?

   ? What information did you lack?

   ? How did you deal with the lack of data?

   ? What assumptions did you make? What did you base your assumptions on?

3. Encourage participants to report their findings to each other, by asking questions such as the following:

   ? What methods did you use to verify your data?

   ? What were the main signs of vulnerability ?

   ? What assessment tools did you use?

   ? What did you learn about vulnerabilities and capacities in disaster-prone communities?

   ? It is often said that women, children and the elderly are the most vulnerable. Did the findings of your assessment concur with this?

   ? What are the main causes of vulnerability?

4. Record the main points of the report-backs / discussion on flipchart.

5. Ask participants what they learnt from this activity, with regard to

   (a) the process of conducting vulnerability / capacity assessments

   (b) some general common problems of Southern African communities.

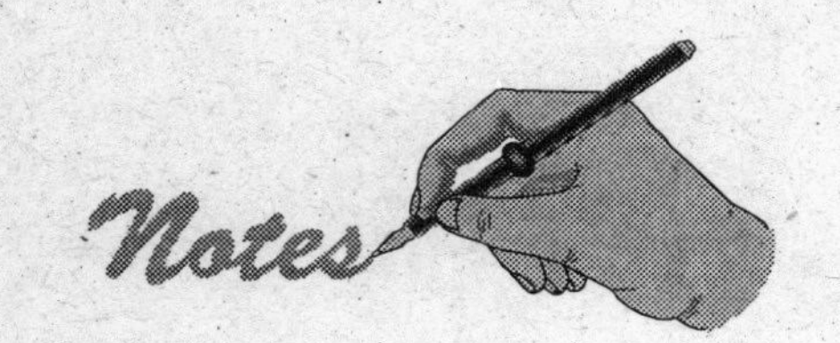
Notes

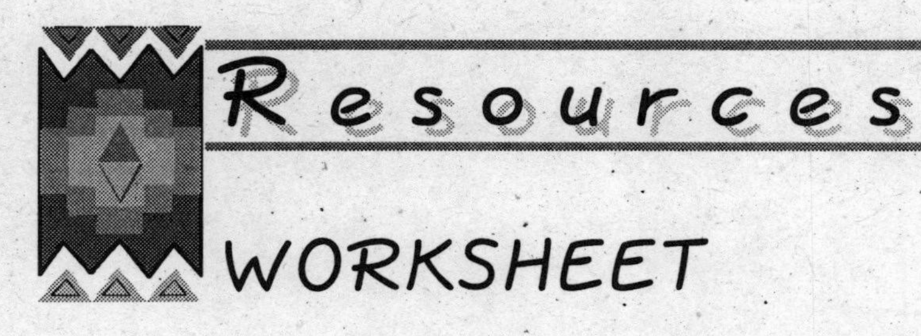

Exploring Vulnerabilities

1. Name your community:

2. Briefly describe the setting - focus on any details that are important to understanding the present situation of this community (geographic, historical, socio-political, demographic, etc).

3. Name the major hazards the community is confronted with.

4. (a) Describe the vulnerabilities (environmental, social, health, organisational, etc) of the community at risk. (Be specific!)

(b) What are the indicators of vulnerability?

5. What resources and capacities do members of the community use in order to cope with the hazard(s)? (Be specific with regard to particular members / sectors of the community.)

6. From your perspective, given the vulnerabilities and capacities outlined under (4)and (5), who is most at risk from the major hazard(s)? Please explain.

What resources are available for reducing risk?

Purpose

This activity aims to help participants identify resources available for short-term risk reduction both at the individual and at community level.

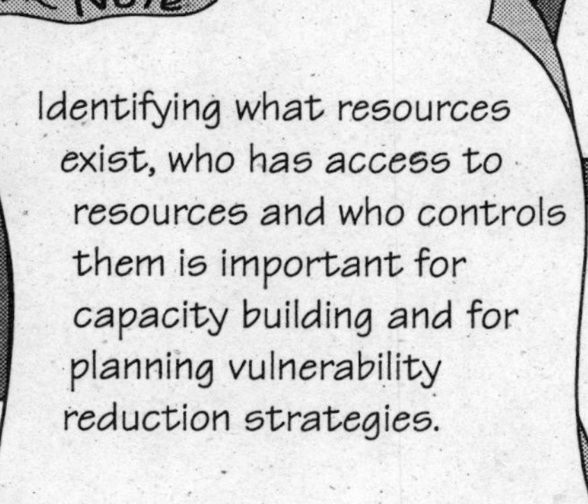

Note
Identifying what resources exist, who has access to resources and who controls them is important for capacity building and for planning vulnerability reduction strategies.

Procedure

In this activity participants use themselves and their own community as a reference point. They brainstorm resources available for reducing risk, categorise them, and discuss them in terms of value, access and control.

Time

- 1 hour

Process

Introduction

1. Outline the purpose and procedure of the activity.

2. Ask participants to do a brief buzz with the person sitting next to them, on the following: what does the term 'resources' suggest to you?

3. In plenary, clarify the term and build a common understanding.

   Suggested Definition

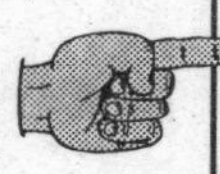

> A resource is something that a person, household, community or country has and that it can use to protect or increase its well-being and wealth. It also denotes a power or ability to do a particular thing.

4. Explain the following :

   In risk reduction, finding out what resources exist allows us to build on what's available. However, community members often consider their particular skills worthless, either because they consider them 'natural' rather than 'learnt', or because they know that in industrial societies such skills have little value.

Sample Story

You may want to think of an example from your experience to illustrate this point. I often tell the following story:

Once in a rural development workshop for at-risk women, I asked an elderly woman what skills she had. She said 'none'. From a previous conversation I knew that she was a broom-maker, and so I pointed out that she did have a skill: she knew how to make brooms. 'Oh', she responded, but that's not a skill, I was born with that'. Only once she discovered that nobody else in the group shared her skill did she acknowledge that, yes, indeed, her ability was a resource that she could use to generate income, and thus reduce her vulnerability.

Participant Action

1. Initiate a brainstorm around 'resources we have'.

   Explain that participants should use themselves (individually and as a group) as a reference.
   Ask for two participants to record what is said, on flipchart; remind them of the rules of a brainstorm: everything anyone says must be listed; at this stage there is no elimination of any suggestion however outrageous it may sound.

2. Ask participants to begin the brainstorm; allow about 5 minutes, or until you have a long list of different suggestions.

3. Ask participants for categories that would allow you to classify the resources; eg. physical (such as 'strength', or 'health'), mental (such as 'patience', or 'compassion'), ability (such as 'writing', or 'knitting') infra-structural (such as 'electricity', or 'water') etc.

4. Collectively, classify the resources listed during the brainstorm according to the categories that have been agreed upon in (3).

Review and Discussion

1. Review the list:
   - ? Are there any obvious omissions? For example, is 'time' included in the list?
   - ? Are there any surprises? What makes those items surprising?
   - ? Are there any disputes about the items on the list?

2. Discuss the list in terms of questions such as the following:
   - ? Which of the resources are personal ones? Which are community ones?
   - ? Are certain resources particular to special individuals / groups of people in a comunity?
   - ? Who controls access to and management of community resources?
   - ? Why do some people not have access to those resources?
   - ? How could even access be arranged?
   - ? How are the resources valued? Which ones have 'status', and on what considerations is the status based?
   - ? How are the resources listed used in our daily lives for food security?
   - ? Are there any resources that we do not draw on? Why? How could they be utilised?
   - ? How would you build on available resources in order to decrease vulnerability?

3. Part of a risk reduction exercise is a process of identifying and assessing available resources. Discuss and outline how participants could use this process in the communities with which they work.

   Point out the need for finding the right word for 'resource', in a community's language.

This activity could be followed by a wealth ranking exercise; in this way participants would increase their awareness of how communities and societies place different values on resources available to them.

Notes

How does gender impact on vulnerability and capacity?

Purpose

Participants will develop a deeper understanding of how drought effects various members of a community differently. They will improve their knowledge of how capacities and vulnerabilities are related to each other.

This article is a case study of how drought impacts on the lives of women within the context of economic, environmental, social and health conditions. It provides useful information about women's coping mechanisms, and illustrates how capacities and vulnerabilities are interlinked. The activity should lead participants to the stage where they can plan disaster reduction strategies that realise the potential of emergency situations for the creation of development programmes.

Procedure

This is a reading and group discussion activity.

Time

- 1½ - 2 hours

Materials

- copies of "The effects of drought on the condition of women" by W. Tichagwa for each participant (see resources)
- pens and paper
- discussion questions written on flipchart

Process

Introduction

1. Outline the purpose and procedure of the activity.

Participant Action

1. Hand out the reading and ask participants to read the article, making notes of key issues such as the strategies for coping with the drought, interrelationship of different factors impacting on food security, etc. (Allow participants 30 minutes for the reading).

2. Ask participants to get into discussion groups of 4-5 people. Reveal the discussion questions written up on newsprint and invite the groups to discuss the article using the questions as a guideline. Point out that there will be no report backs from the discussions.

   The following are sample questions; you may want to develop your own:

   — *women perform particular roles in their families and communities. What are the effects of drought coping mechanisms on these roles?*

   — *how do the activities which women employ as coping mechanisms further increase their vulnerability?*

   — *how could those activities be utilised for boosting womens' capacities and decreasing the risks imposed by the drought?*

   (You have 45 minutes for this discussion)

Review and Discussion

1. In plenary take up any points that might need clarification.

2. Ask participants to give other examples of coping strategies that people in their own communities use in drought situations.

3. Analyse briefly the potential impact of those strategies within the framework of vulnerabilities / capacities.

4. Refer to the 'policy and planning implications' in the article for examples of how emergency situations often present possibilities for introducing mitigating interventions.

5. Request participants to get back into their discussion groups and introduce the following task:

   — *list further suggestions of how short-term relief could be combined with long-term development programmes.*

   Write your ideas on flipchart.

   (You have about 10 minutes for this task).

6. Briefly ask for report-backs and review the suggestions in the light of the following question:

   Do they strengthen rather than further burden women's position?

7. Ask a participant to summarise the key issues from the session.

Resources

READING

The effects of drought on the condition of women[1]

Wilfred Tichagwa

This paper, looks at the real or potential impacts of drought on the material conditions of life as they affect rural women. The rationale for focusing on rural women is two-fold. First, rural women are the backbone of the rural economy. Any change in the condition of women which will affect their performance in economic activities will inevitably affect the performance of the rural economy as a whole. The second point is that at household level, women are to a large extent responsible for food provision and the overall survival strategy of the family. The effects of drought are therefore important to the extent that they undermine the women's efforts to fend for their families.

The following discussion focuses on the economic, environmental, social and health impacts of drought on women.

Economic impacts

As shown in Figure 1, crop failures caused by drought result in food deficits in terms of the household's needs. Also there will be little or no crop surplus for sale, therefore income from his source is reduced or even wiped out completely. It will not be possible therefore for women to buy food when stocks are depleted.

A further possible consequence is increased male labour migration. The women remaining at home end up with an increased number of tasks as they must now do the work for which the migrant males were responible. Where the men stay away for extended periods, the increased burden of women could result in reduced agricultural productivity. Figure 1 shows that this could lead to vicious circles in which reduced productivity leads to food deficits and reduced income form the sale of reduced surpluses. This in turn could lead to a reinforcement of the male labour migration as part of the solution in the household's needs for food and cash income.

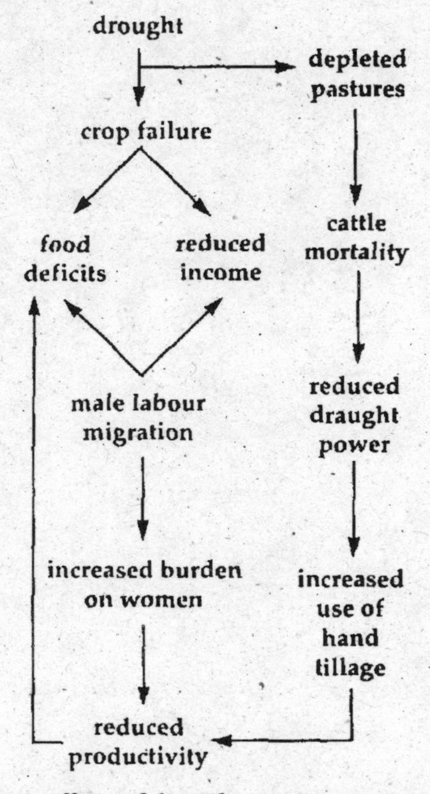

Figure 1: Effects of drought on women's agricultural production and marketing

The right-hand part of figure 1 shows another effect of drought on agricultural production. Depletion of pastures leads to high mortality of cattle. The resultant reduction in a power means that women have to till the fields by hand which is back-breaking work; and women are often forced to reduce the area they cultivate.

At national level, the traditional response to drought is the food-for-work or money-for-work programme. Often, the food-for-work projects rely mostlyon women's labour because women are seen as the providers of food. Men tend to prefer money-for-work projects as these are seen as substitutes for paid employment, traditionally regarded as men's area. Drought-relief programmes thus tend to reinforce the traditional division of labour which places on women a heavy responsibility for reproductive work.

Finally, it is obvious that hunger and malnutrition will leave people in a weak physical condition. There will therefore be a drop in productivity until the nutrition status of the labour force is restored to previous levels. Child labour, in particular, will be greatly reduced. Since children's work is usually undertaken as a helping hand to mothers, then the weaker the children, the more work their mothers must take back from the children. Such tasks as fetching firewood, washing dishes, fetching water, cleaning the house and yard, laundry, gardening, food preparation, and child minding, will all have to be done by the mothers on their own. This, in turn, will have adverse effects on agricultural production.

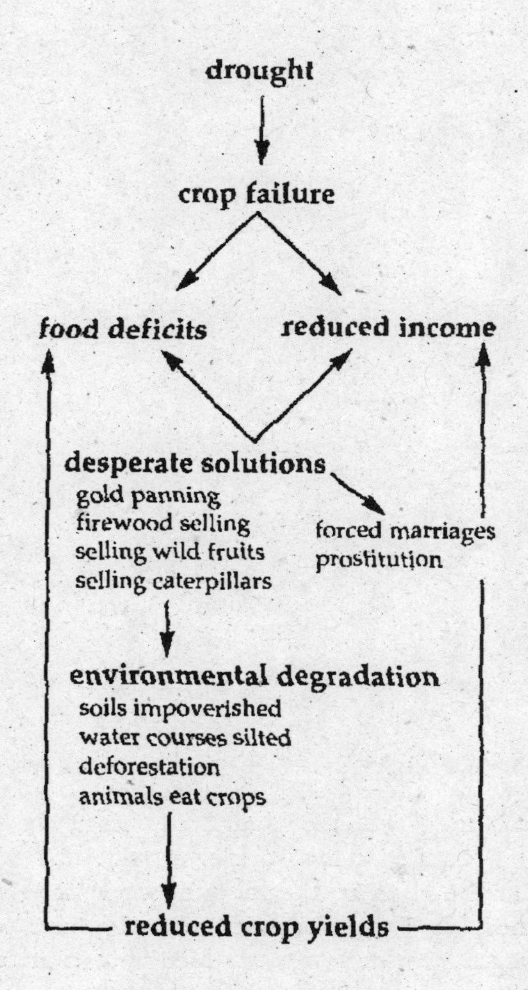

Figure 2: Environmental and social consequences of drought

Environmental impacts

In Figure 2 it can be seen that in some instances women will resort to desperate measures to avert the crisis of hunger in the family. For example, gold-panning in river banks is on the increase in many rural areas. Also the collection of wild fruit and caterpillars for eating and selling has increased in the last decade. These activities have enabled families to supplement their diets in the times of drought. However, they can be dangerous, as when holes dug into river banks cave in and entomb gold-panners; and they can create environmental problems, such as the extensive silting of water courses resulting from indiscriminate gold-panning.

In Matebeleland South, collection of caterpillars has escalated as a consequence not only of recurrent drought but also of the commercial processing and packaging of caterpillars by some big food companies. Villagers have been felling trees for easy access to the caterpillars. The Mopani tree on which caterpillars are found is a slow growing hardwood, therefore regeneration will take decades. The excessive harvesting of caterpillars threatens them with extinction.

The collection of wild fruit for sale is also reaching alarming proportions. In the area around Domboshawa, the writer witnessed the felling of trees so that fruit could be collected. The fruits were not ripe and therefore could not be shaken down from the high branches, and as the tree has a weak bark and a rather brittle trunk, it is unsafe to climb. The solution is to cut the tree down. As a result, this fruit tree is becoming extinct in some areas.

A group of women were lamenting the likely consequences of tree felling to collect upripe fruits. Firstly, there will be less fruit the following year. Secondly, harvesting both ripe and upripe fruit leaves nothing for the baboons and monkeys to eat at a time when farmers are planting maize seeds. As the animals are capable of uncovering and eating a large proportion of the maize seed, there was a fear that this will result in very low germination rates and reduced crop yields.

Tree felling for firewood for hungry people to sell is also a serious threat to the environment, and has resulted in extensive deforestation in some communal areas and resettlement areas. Denuding the land of trees makes it extremely vulnerable to erosion and gully formation.

These desperate solutions to the problem of recurrent food deficits will result in environmental degradation so that in future women will have great difficulty in finding firewood, and timber for building purposes. Denuding the land of its forest cover will result in impoverished soils and consequently poor crop yields. Excessive tree-felling could result in reduced evapo-transpiration, a vital element in the water cycle.

Social impacts

Recurrent drought has in many instances resulted in increased labour migration. In the rural areas, many households are now female-headed. Where the migrant husbands do not remit cash regularly, the wives have a difficult time tying to run the affairs of the household.

In some cases, the trappings of city/town life have turned the otherwise seasonal migrant workers into permanent urban dwellers, with a second wife or more in their second homes. A husband in town may even require his wife in the rural areas to send food to him and his other wife after the harvest, and this practice increases in drought periods.

Another unpleasant consequence of chronic food deficits has been forced marriages for girls so that their parents can survive on the *lobola* (bride price) paid by her husband. Wives of relatively prosperous men face the danger of finding themselves in a polygamous marriage. In the peasant sector, the amount of a man's wealth often correlates directly with the number of wives he has. Polygamy gives a man a large reserve of unpaid labour and makes him the envy of other men. Thus, for a hunger-stricken and desperate family - usually it is the father who is anxious to marry off a daughter as the means of survival - there is no shortage of would-be-polygamists. For a young girl, it is a high price to pay for the benefit of others, not herself. It must also be a cause of great sorrow and anguish for the mother as she watches helplessly.

Prostitution is increasing among women, particularly unmarried mothers, in an effort to earn income to support their family. Prostitution carries with it the ever present danger of sexually-transmitted diseases, including HIV/AIDS. Other women face the danger of infection by promiscuous husbands and/or boyfriends.

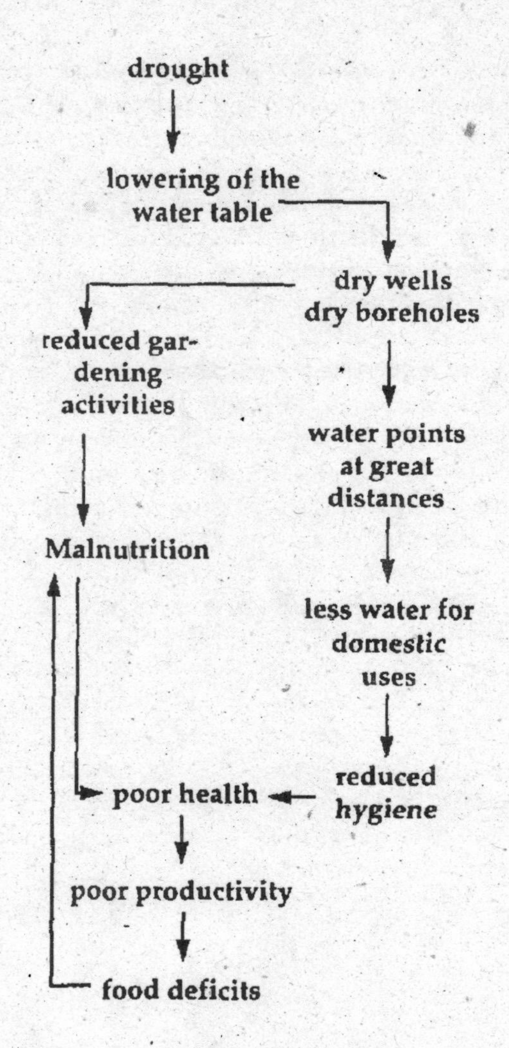

Figure 3: Health impacts of drought

Health impacts

Figure 3 illustrates a situation where progressive lowering of the water table due to drought leads to the reduction of gardening activities as wells and boreholes dry up. This has adverse consequences for the nutritional status of the family.

Another consequence is that access to clean water for domestic purposes will be difficult as water points get farther and farther away. Women will not have time to collect sufficient water to meet the family's needs. The consequent reduction in domestic and personal hygiene has obvious implications for health. Poor health, in turn, result in reduced agricultural productivity and consequent food shortage.

The increased mortality of domestic animals means that there will be less milk and meat for the family, with very serious effects on nutrition and health. The loss of draught power and resultant increase in the exhausing work of hand-tillage ultimately takes a toll on the women' health. It is likely that some women's life expectancy has been adversely affected by drought.

Food shortages also pose health risks to pregnant mothers and their unborn babies; and poor nutrition in the early years can have life-long consequences for young girls. Girls who are healthy and well-fed during their own childhood and teenage years have fewer problems in pregnancy and at childbirth.

It should also be noted that rural hospitals do not provide food to mothers-in-waiting and their attendants. Because of food shortages at home, these mothers cannot take much food with them they go to hospital to seek antenatal care, and many women do not go to the hospital at all, for this reason. The inevitable consequence will be increased levels of maternal mortality as the combination of under-nourishment and inadequate antenatal care takes it toll.

Drought-induced poverty also results in reduced access to medical care. Cost recovery measures were introduced for health services as part of the Economic Reform Programme. Many women and their children cannot now afford medical care because their resources of income - crop surpluses - were wiped out by the drought.

Policy and planning implications

Drought relief and rehabilitation of food production must respond to the needs of women farmers in a manner that increases their capacity to withstand the effects of drought, at the same time reducing the burden of reproductive work. Future relief and rehabilitation programmes should aim at:

1. Meeting immediate food needs

2. Strengthening women's role as farmers in their own right.

3. Introducing environmentally sustainable long-term solutions to food deficits.

Current programmes are heavily biased towards short-term solutions such as food handouts, work projects and supplementary feeding schemes. Such interventions may immediately improve the material conditions of life in the existing drought situation, but do not empower women to fend for themselves in a future drought.

In future, reform programmes should be introduced alongside the traditional solutions to drought-related problems, with a view to improving the positions of them as farmers. Women should enjoy the same rights to arable land as men, the same access to extension services and agricultural credit, and equal control of agricultural produce and income. This will increase women's ability to plan for and maintain greater food self-sufficiency at household level, with cumulative effects at local level. While nothing can prevent drought, such reforms would enable women to plan for food production and make provision for a possible drought in the following season.

The effects of drought can be reduced through environmental-protection programmes, such as controlled-grazing schemes, reforestation projects, gully-reclamation projects, construction of silt traps, and construction of dams for irrigation purposes. Such programmes will reduce overgrazing and denudation of the land as well as improve the organic content and moisture-retention capacity of the soil, and should be in the form of on-going programmes rather than food-for-work projects in drought periods.

The combined effects of the socio-economic reforms and environmental protection programmes proposed here would improve women's socio-economic status and enhance their agricultural productivity. In turn, the role of women farmers in ensuring food self-sufficiency and food security would be stregthened, to the benefit of their families and society as a whole.

It would be wise to maintain a reasonable level of preparedness at national level to respond to severe drought. We now know that drought is a recurrent phenomenon. There is therefore no need to be caught unprepared. Both government and NGOs should allocate resources for appropriate contingency plans. Such plans would include a stand-by Drought Relief Fund or a Foodgrains

Bank, or both. To facilitate effective food-distribution, a system of roads and convenient storage and distribution points should be established. This would avoid the costs of foreign loans to fund drought-relief operations in a country that is quite capable of feeding itself; or the embarrassing situation where existing food reserves cannot be delivered to starving people because of the poor road network.

Above all, contingency drought-relief plans should recognise the exisiting burden to reproductive work on women, rather than assume that women have an unlimited capacity to sustain this burden.

In short, what is being advocated is gender-sensitive forward planning for the prevention of food deficits in the event of drought. Where such food deficits are unavoidable, food distribution systems should be sensitive to women's existing burdens, and should not lose sight of the need to empower women to produce their own food as far as possible in future.

This paper was presented at a workshop on drought organised by the Zimbabwe Women's Resource Centre and Network in December 1992. Published in Focus on Gender 2 (1), 1994.

How do we assess household food security?

Purpose

Participants practise how to use household food security questionnaires, and process the information to assess vulnerability.

In order to target individuals and households most vulnerable to food insecurity development workers must assess individual households. Often this is done by questionnaire. In the process of administering assessment questionnaires participants critically examine the questionnaire as a tool for conducting a food security assessment.

Procedure

This is a role play in which participants assume the roles of development workers conducting assessments, or household representatives.

Time

- 2 hours

Materials

- two sets of household briefing cards (see resources)
- sample questionnaires for each group of participants (see resources)

Process

Introduction

1. Introduce the activity by outlining the purpose and procedure.

2. Explain that this exercise focuses on household assessments; ask participants to respond to the following questions:

   ? What type of questions would you ask in order to assess food security?

   ? What signs of food security / insecurity would you look for?

   ? How would you go about testing whether your signs and questions are right?

   Record suggestions on flipchart.

3. Distribute copies of the 'food security assessment' questionnaire (see resources). Briefly, check for understanding of the questions. Compare them with the sample questions on the flipchart: there should be an overlap. Briefly discuss differences and similarities. Point out that the following activity asks particpants to test the questionnaire.

Participant Action

1. Ask participants to get into two groups: both groups will be made up of household representatives, two community development workers, and an observer. Participants may hold more than one household briefing card. Point out that the groups will work independently from each other.

2. Appoint 'development workers' and 'observers' and distribute all the household briefing cards to the remaining participants. Ask participants to read through their cards.

3. Give the following instructions to household representatives and development workers:

HOUSEHOLD REPRESENTATIVES: Read through your household briefing cards and make sure you understand them. Think yourselves into your roles: what are your household's strengths and weaknesses? Construct a profile of the household in order to fill in information not given on your briefing card. Prepare to be questioned.

DEVELOPMENT WORKERS: plan a course of action: how will you go about interviewing the households, and recording their responses?

4. Give the following instructions to observers:

> OBSERVERS: observe the process of interviewing: focus on both the process and the actual questions. Note down problems or difficulties during the interviews. Prepare to report-back..
>
> You have 5 minutes to prepare and approximately 25 minutes to conduct the interviews.

5. Monitor the process and assist where necessary.

Review and Discussion

1. Call participants together and initiate a process of unpacking the role play.
2. De-brief the players and review the process of interviewing:
   - **?** What resources did participants draw on in order to play their roles?
   - **?** How did they feel about being interviewed?
   - **?** How did the development workers feel about asking questions?
   - **?** What resources did they draw on in order to facilitate the process of interviewing?
   - **?** Ask observers to report their observations regarding the interviewing process.
3. Facilitate a discussion on 'helpful things to do when interviewing vulnerable households'.

   Summarise the main recommendations that arise from the discussion on flipchart.

Example:

> HELPFUL THINGS TO DO WHEN INTERVIEWING...
>
> — consult with community representatives / leaders in advance of the assessment process
>
> — ensure you know culturally appropriate forms of address and greeting
>
> — record data together with informant
>
> — make your own observations about the well-being of the household

4. Focus on the questionnaires and ask questions such as the following:
   - Which questions worked well / generated useful information
   - Which questions were problematic?
   - Which questions generated the most useful data about household food insecurity? Why / How?
   - How do you know whether responses given are an accurate reflection of the respondant's circumstances?
   - Given the data collected - which households would you rate as being most vulnerable? What are the criteria used for the decision?

5. Sum up what has been learnt about assessing food security.

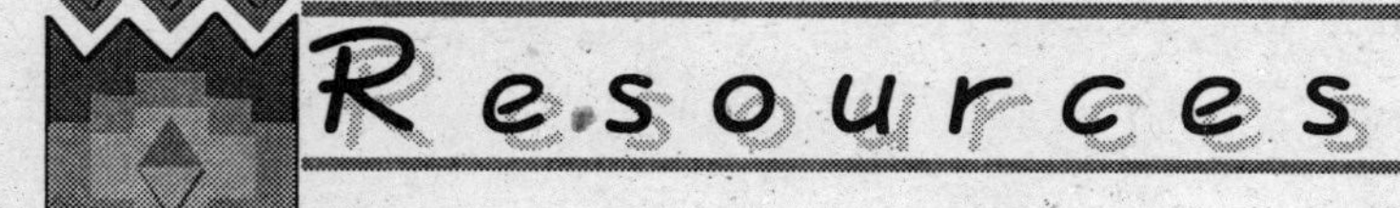

ROLE PLAY

Household briefing cards.

(1) You are a woman in her mid-thirties. There are 7 members in your household: your four children (two under 5), one of whom has a disability due to polio, your sister and her teenage daughter. Since your husband abandoned you three years ago you have been the head of the household. You have two small plots of land and in a good year you manage to grow enough maize, beans and groundnuts to see you through the year. You also raise chickens. Your sister contributes financially, when she manages to find piece-work. Your oldest son works once a week as a gardener; his employers give him old clothes. You attended school for 7 years.

(2) You are the head of a household of 6: your wife, her mother, your younger brother, and two children. You have a basic school education and as a young man you worked as a mechanic apprentice; now you are unemployed. You like listening to the radio; you sing in a choir; you own some cattle and small livestock. Your homestead is situated next to the river and you use most of your land to grow vegetables.

(3) You and your three children live in a small mud house in the compound of your brother-in-law; you do not own any land. Your husband works in the mines in South Africa and only comes home once a year. Your eldest daughter collects wood and sells it; your son looks after his uncle's goats and sheep. Your baby suffers frequently from respiratory infections. You have to walk far in order to draw water. Your prize possession is a sewing machine.

(4) Your most important possession is an old car which you picked up with the help of your employer in town. You are currently at home on a visit; your second wife just gave birth to twins. You married her when your first wife died from AIDS. You are now a father of 4 boys and 3 girls; other members of your household include your cousin's daughter who helps out in the house, and your aged sister who suffers from severe arthritis. You have recently erected a new extension to your house and built a toilet. You own 4 large fields, including some fruit trees.

(5) You are a widow of forty; you have three grown-up children who have moved to their own homes. Your one grand-child lives with you because there is a school near your house. Your husband used to grow enough maize and vegetables to feed you all; he established a rainwater tank and you still have farming implements for three people. You used to have livestock but now you can only manage a few chickens and a pig. Your children come to visit every other month; they give you some money so that you can afford to pay medical bills, or buy essentials.

(6) You are a pensioner looking after 5 grand-children, whose mother lives and works in town; there is no trace of her husband. Three of the children are under five years old, they were brought to you severely malnourished. You know the water from the river is bad, and you have a constant fear of infections leading to diarrhoea. You manage to grow a small amount of vegetables, and you have a dairy cow. You are much respected in the community for your knowledge of health care.

(7) You are a widower in his early fifties; you moved back to the rural areas after surviving a politically motivated attack on your life that left you disabled. You managed to get a loan in order to buy a tractor which you hire out. Your daughter and her two school-going children live with you. She sells second-hand clothing. You live with relatives and own no land. You dream of one day building your own homestead with a house for your daughter and her family.

(8) You are the mother of 6 children, two under-5, three of school-going age and a 16 year old daughter who is pregnant. Your second-born is sickly and painfully thin. Your husband owns large fields and grows mainly maize; this years' harvest was poor. He is often drunk and abusive. You know that he sells maize stocks for beer. You sell home-grown vegetables as your main source of income and you are an active member of the school committee.

(9) You and your wife work hard on the fields in order to grow a variety of cereals and vegetables; your one son tends the 5 heads of cattle while the other still goes to school. Your daughter fetches water and firewood and helps with the baby, but you insist that she must attend school as long as possible. Your young sister and brother help you in the fields; they also work for other members of the community. You have a radio and a bicycle. You are very worried about the ongoing drought because you live far away from the spring.

(10) You are an old woman, tired from a long life of hard labour. You live in a small house with your husband who is sickly and can no longer work. You have not heard from your sons and daughter for a long time; a young girl, the daughter of a cousin, looks after you.

Questionnaire for Food Security Assessment

1. Identification of homestead / household

2. Indicate the following about resident household members (including yourself)

	self	1	2	3	4	5	6	7	8	9	10
a) Age											
b) Sex											
c) Marital Status											
d) Relationship to head of HH											
e) Years of school											
f) Main occupation											
g) Amount of money contributed to household / month											
h) Fitness for physical work Y / N											
i) Type of work											
j) Availability for work / week											
k) Is s(he) involved in other community development projects? Describe.											

3(a) Information of children under 5 years of age (use code)

Member no. from 2, above	Age in months	Illnesses in the last 2 weeks

CODE

1. Respiratory
2. Intestinal
3. Malnutrition
4. Skin & tissue
5. Viral disease
6. Other (specify)

(b) Number of pre-school children ____________

(c) Number of school age children in the household ________

(d) Number of children in school ______________

4. Number and types of structures in household:

Rooms _______ Building material _______

5(a) Livestock owned:

i) Cattle _______ ii) Goats and sheep_______ iii) Chickens _______
iv) Pigs _______ iv) Other (specify) _______

(b) Household possessions

Item	No.	Item	No.	Item	No.
Plough		Sewing / knitting machine		Coal / gas stove	
Hoe		Television		Tractor	
Spade		Radio		Vehicle	
Other		Other		Bicycle	

(c) Ownership of fields or land:

Field no.	Planted last year yes / no	Size	Planted this year yes / no	Size
Field 1	Y N		Y N	
Field 2	Y N		Y N	
Field 3	Y N		Y N	
Field 4	Y N		Y N	

6. Horticulture: number of fruit trees _______
Number of garden crops _______

7. Household Facilities (circle if used now; tick if exists but not used):

WATER SUPPLY	FUEL	TOILET
Piped on site	Electric	Flush toilet
Private boreholes	Gas	VIP (NRSP)
Communal piped	Coal	Other
Communal handpump	Paraffin	Latrine
Covered spring	Charcoal	Bucket
Other spring	Wood	Fly screen
Rainwater tank	Dung	None
River	Crop residue	
Dam	Shrub/weeds	

Time to collect: _______ _________

8 (a) Does anyone in the household have a bank account now?

1. Yes 2. No 3. Amount: _______

(b) What is the main source of income for this household? ____________________

(c) Where / how else do you get income? ____________________

(d) Expenditures in last week:
Maize ____ Wild spinach ____ Sugar ____ Fuel ____ Water ____ Transport ____

(e) Expenditure in last 6 months:
School fees ____ Health ____ Building ____Furniture ____ Animals ____Clothing _____

(f) Farming expenditure this season:
Fertilizer ____ Seed ____ Pesticide ____ Ploughing ____ Labour ____

9 (a) When are the months when it is most difficult for you to feed your family?
Beginning month ____________ to month ____________

(b) In a good harvest year do you get enough food to feed your family? Yes, No.

(c) Food and crop grown this year and condition of crop

Field no.	Anticipated harvest		Condition of crop Good / Fair / Poor
	Month	No. of Units	
Maize			
Beans			
Sorghum			
Wheat			
Other (specify)			

(d) Cereal stocks in household (bags and tins)
Maize ________ Sorghum ________ Wheat ________ (units)

(e) Please describe what household members have eaten during the past 24 hours:

	Morning	Afternoon	Evening
Infants			
School-age children			
Adults			

10. General Comments: (housing / well-being etc) ____________________

____

____

____

How do we target households most at risk?[1]

Purpose

Participants improve their understanding of the pros and cons of using different approaches and criteria in targetting at-risk households.

Note

Participants are asked to make decisions about who should be targetted for a risk reduction programme in a drought-stricken area.

Procedure

This activity is a simulation in which participants assume the role of households in a small village.

Time

- 2 hours

Materials

- observer instructions (see resources)
- household information cards (see resources)

Process

Introduction

1. Introduce the activity by outlining its purpose and procedure.

2. Explain that this activity builds on programmes in which the most vulnerable households in a community are targetted for relief or development action. Often the beneficiaries of such action are not people most at risk. This happens for two main reasons: firstly, because the process of targetting is difficult, and secondly, the decision-makers within communities are rarely the poorest members of the community.

3. Point out that this activity assumes a clear understanding of concepts like 'targetting', 'community', 'household' and 'wealth'.

Agreeing on common terms

1. Ask participants to define 'community' and 'household'.

2. Write the key terms on flipchart.

3. Discuss some of the problems that might arise when defining 'community' and 'household'; issues that might arise include the following:

   ? How are the boundaries of a 'community' defined? What are the markers or criteria used to define the boundaries? How does the community signal it's boundaries to the outside?

   ? How do different communities define 'household'? Who belongs to the unit and who is excluded? What are the criteria used? How does the delineation relate to the question of head of household? Whose name is given to the household?

4. Work towards an agreed use of terminology.

Participant Action

1. Introduce the role play by describing the process:

   — most participants will get one or more household information cards; three participants will be observers with specific instructions; you will act as an NGO field officer

   — participants will work for 30 minutes as a group making a decision about who to target for a risk reduction programme

   — in plenary, you will unpack the activity and discuss implications

2. Ask for three volunteers to act as observers; hand them the observer instruction sheets.

3. Distribute the household information cards to the rest of the participants. Point out that participants will act as representatives of those households.

4. Point out that the information on the cards is based on real data collected from a village comprising 110 households in northern Ghana. Go through one of the cards and explain what the information means.

5. Introduce yourself as the local field officer of an NGO.

6. Call participants together and inform them that this is an important meeting of household representatives.

   You will address this meeting.

7. Make the following announcement (you may want to write the key terms on newsprint and point to them as you speak):

> "I have come to assist this community. My organisation is implementing a risk reduction scheme in the region. We note that in the recent drought you have had devastating losses of livestock. We offer you the following: **interest free loans of 50.000 Cedis, repayable over four years, available to half the households in this community, for purchase of livestock** to replace those lost in the drought. We will organise the buying of the livestock from outside the drought-affected region.
>
> It is up to you to decide which households are to be eligible for the loans. However, you have only 30 minutes to make your decision. If you fail to agree within this time no loans will be disbursed."

8. Ask whether participants / household members have understood the announcement and instruction. Repeat the key points, if necessary.

9. Withdraw from the process and allow participants to find their own way of dealing with the task. Let the session run for more than 30 minutes if this will help the participants to arrive at a decision.

Review and Discussion

1. Thank all participants for playing and initiate a plenary discussion; ask questions such as the following:

   - **?** Describe how you found the process: How did you feel? What did you think?
   - **?** What were the main problems you faced?
   - **?** How did you work together as a group?
   - **?** Did all participants have an equal say in the process? Did each voice carry equal weight?
   - **?** What were the main considerations and preoccupations on which you spent time?
   - **?** How were decisions made? Who made them?

2. Ask observers to report their responses to questions (i) assumptions made, and (ii) process of cooperation and decision-making as a community.

3. Ask participants to focus on the criteria for selecting half of the households: what selection criteria did they use? Record them on newsprint. Check with observers whether this list is accurate and complete.

4. Ask participants to examine the advantages and disadvantages of the various criteria.

   - **?** How did they go about eliminating them?
   - **?** What are the crucial questions to ask in a wealth assessment of this kind?

   Write key responses on newsprint.

5. Ask participants to consider the decision-making process.

   ? Was there a problem about this being a mixed man/woman group?

   ? How does this relate to the field: who (generally) does most of the talking in a community meeting such as this?

   ? What is the impact of who speaks on the decisions reached?

6. Ask participants to reflect on the role of the donor agency. Ask questions such as the following:

   ? What are the strengths and weaknesses of this form of targeted credit?

   ? In selecting criteria for the allocation of funds, which considerations are more important: the criteria of the donors, or the sense of fairness in the community?

   ? What happens if a sense of fairness amongst the community is ignored?

   ? What is the difference between a welfare approach and a development approach to risk reduction?

7. Sum up by asking each participant to make a brief statement that summarises what s/he learnt in this activity.

1. This activity is based on a role play prepared by Food Studies Group, Queen Elisabeth House, 21 St Giles, Oxford OX1 3LA, UK. It was facilitated and tested in the SADMTP course, in April 1995, by Graham Eele.

Notes

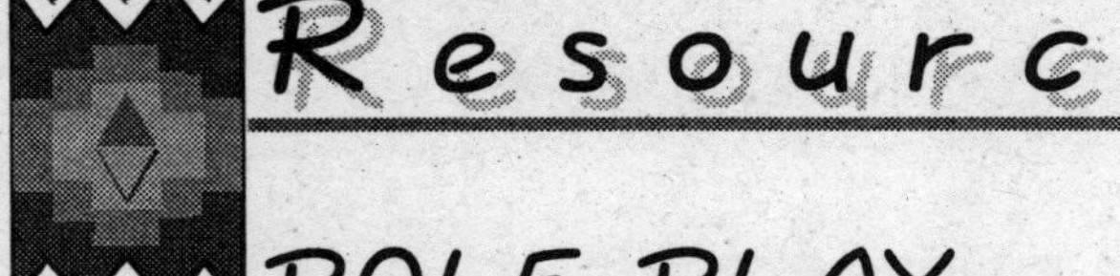

ROLE PLAY

Observer instructions

Please observe the community meeting and take notes of the following:

(1) assumptions made by the heads of household

(2) the process of discussion: who speaks? Who dominates? Who is silenced?

(3) the selection criteria suggested

(4) the procedure for eliminating criteria and making decisions

You will be asked to report your observations.

Household Information Cards

HOUSEHOLD INFORMATION CARDS											HOUSEHOLD NO. 1.34.2				
ANIMALS						FARM TOOLS				CONSUMER GOODS					TOTAL ASSET VALUE
BULLS	COWS	GOATS	SHEEP	FOWLS	VALUE	PLOUGH	CUTLASS	HOES	VALUE	BIKES	RADIOS	LAMPS	ROOFS	VALUE	
0	0	0	0	2	1 200	0	2	5	4 000	0	0	1	0	7	5 900

INCOME				FOOD SECURITY				HOUSEHOLD SIZE		
CROP SALES	ASSET SALES	OFF FARM INCOME	TOTAL INCOME	BOUGHT GRAIN	RATIONED FOOD	MEALS/DAY		ADULTS	CHILD	60+
						ADULT	CHILD			
7 800	1 120	12 000	20 920	YES	YES	1	2	2	4	0

HOUSEHOLD INFORMATION CARDS											HOUSEHOLD NO. 1.37.1				
ANIMALS						FARM TOOLS				CONSUMER GOODS					TOTAL ASSET VALUE
BULLS	COWS	GOATS	SHEEP	FOWLS	VALUE	PLOUGH	CUTLASS	HOES	VALUE	BIKES	RADIOS	LAMPS	ROOFS	VALUE	
0	0	2	0	2	8 200	0	3	4	3 900	0	0	1	0	700	12 800

INCOME				FOOD SECURITY				HOUSEHOLD SIZE		
CROP SALES	ASSET SALES	OFF FARM INCOME	TOTAL INCOME	BOUGHT GRAIN	RATIONED FOOD	MEALS/DAY		ADULTS	CHILD	60+
						ADULT	CHILD			
0	18 200	2 000	20 200	YES	YES	1	1	2	3	0

HOUSEHOLD INFORMATION CARDS											HOUSEHOLD NO. 1.38.4				
ANIMALS						FARM TOOLS				CONSUMER GOODS					TOTAL ASSET VALUE
BULLS	COWS	GOATS	SHEEP	FOWLS	VALUE	PLOUGH	CUTLASS	HOES	VALUE	BIKES	RADIOS	LAMPS	ROOFS	VALUE	
0	0	0	0	3	1 800	0	2	5	4 000	0	0	1	0	700	6 500

INCOME				FOOD SECURITY				HOUSEHOLD SIZE		
CROP SALES	ASSET SALES	OFF FARM INCOME	TOTAL INCOME	BOUGHT GRAIN	RATIONED FOOD	MEALS/DAY		ADULTS	CHILD	60+
						ADULT	CHILD			
21 800	0	6 000	27 800	YES	YES	2	2	3	4	0

HOUSEHOLD INFORMATION CARDS											HOUSEHOLD NO. 1.42.1				
ANIMALS						FARM TOOLS				CONSUMER GOODS					TOTAL ASSET VALUE
BULLS	COWS	GOATS	SHEEP	FOWLS	VALUE	PLOUGH	CUTLASS	HOES	VALUE	BIKES	RADIOS	LAMPS	ROOFS	VALUE	
4	4	5	5	14	300 900	2	7	11	60 100	2	2	10	0	232 000	593 000

INCOME				FOOD SECURITY				HOUSEHOLD SIZE		
CROP SALES	ASSET SALES	OFF FARM INCOME	TOTAL INCOME	BOUGHT GRAIN	RATIONED FOOD	MEALS/DAY		ADULTS	CHILD	60+
						ADULT	CHILD			
41 200	10 400	448 000	499 600	NO	NO	3	3	10	7	1

HOUSEHOLD INFORMATION CARDS

HOUSEHOLD NO. 1.10.2

ANIMALS						FARM TOOLS				CONSUMER GOODS					TOTAL ASSET VALUE
BULLS	COWS	GOATS	SHEEP	FOWLS	VALUE	PLOUGH	CUTLASS	HOES	VALUE	BIKES	RADIOS	LAMPS	ROOFS	VALUE	
0	0	5	60	25	50 500	0	45	95	7 400	1	2	5	1	60 500	118 400

INCOME				FOOD SECURITY				HOUSEHOLD SIZE		
CROP SALES	ASSET SALES	OFF FARM INCOME	TOTAL INCOME	BOUGHT GRAIN	RATIONED FOOD	MEALS/DAY		ADULTS	CHILD	60+
						ADULT	CHILD			
102 400	0	190 200	292 600	YES	NO	2	3	3	4	0

HOUSEHOLD INFORMATION CARDS

HOUSEHOLD NO. 1.12.2

ANIMALS						FARM TOOLS				CONSUMER GOODS					TOTAL ASSET VALUE
BULLS	COWS	GOATS	SHEEP	FOWLS	VALUE	PLOUGH	CUTLASS	HOES	VALUE	BIKES	RADIOS	LAMPS	ROOFS	VALUE	
0	0	1	2	10	15 500	0	2	5	4 000	1	0	4	3	65 800	85 300

INCOME				FOOD SECURITY				HOUSEHOLD SIZE		
CROP SALES	ASSET SALES	OFF FARM INCOME	TOTAL INCOME	BOUGHT GRAIN	RATIONED FOOD	MEALS/DAY		ADULTS	CHILD	60+
						ADULT	CHILD			
23 400	11 000	1 000	35 400	YES	YES	2	3	3	8	0

HOUSEHOLD INFORMATION CARDS

HOUSEHOLD NO. 1.12.3

ANIMALS						FARM TOOLS				CONSUMER GOODS					TOTAL ASSET VALUE
BULLS	COWS	GOATS	SHEEP	FOWLS	VALUE	PLOUGH	CUTLASS	HOES	VALUE	BIKES	RADIOS	LAMPS	ROOFS	VALUE	
0	0	3	1	7	17 700	0	2	6	4 600	0	0	3	2	32 100	54 400

INCOME				FOOD SECURITY				HOUSEHOLD SIZE		
CROP SALES	ASSET SALES	OFF FARM INCOME	TOTAL INCOME	BOUGHT GRAIN	RATIONED FOOD	MEALS/DAY		ADULTS	CHILD	60+
						ADULT	CHILD			
13 000	1 400	30 000	44 400	YES	YES	2	3	3	8	0

HOUSEHOLD INFORMATION CARDS

HOUSEHOLD NO. 1.13.1

ANIMALS						FARM TOOLS				CONSUMER GOODS					TOTAL ASSET VALUE
BULLS	COWS	GOATS	SHEEP	FOWLS	VALUE	PLOUGH	CUTLASS	HOES	VALUE	BIKES	RADIOS	LAMPS	ROOFS	VALUE	
0	0	3	1	3	15 300	0	2	5	4 000	0	0	2	0	1 400	20 700

INCOME				FOOD SECURITY				HOUSEHOLD SIZE		
CROP SALES	ASSET SALES	OFF FARM INCOME	TOTAL INCOME	BOUGHT GRAIN	RATIONED FOOD	MEALS/DAY		ADULTS	CHILD	60+
						ADULT	CHILD			
15 600	3 200	0	18 800	YES	YES	2	3	2	3	0

HOUSEHOLD INFORMATION CARDS — HOUSEHOLD NO. 1.44.1

ANIMALS						FARM TOOLS				CONSUMER GOODS					TOTAL ASSET VALUE
BULLS	COWS	GOATS	SHEEP	FOWLS	VALUE	PLOUGH	CUTLASS	HOES	VALUE	BIKES	RADIOS	LAMPS	ROOFS	VALUE	
3	5	10	21	19	344,400	2	4	20	64 000	2	1	17	12	239,900	648,300

INCOME				FOOD SECURITY				HOUSEHOLD SIZE		
CROP SALES	ASSET SALES	OFF FARM INCOME	TOTAL INCOME	BOUGHT GRAIN	RATIONED FOOD	MEALS/DAY ADULT	MEALS/DAY CHILD	ADULTS	CHILD	60+
34 700	0	228 000	262 700	NO	NO	3	4	13	22	1

HOUSEHOLD INFORMATION CARDS — HOUSEHOLD NO. 1.46.1

ANIMALS						FARM TOOLS				CONSUMER GOODS					TOTAL ASSET VALUE
BULLS	COWS	GOATS	SHEEP	FOWLS	VALUE	PLOUGH	CUTLASS	HOES	VALUE	BIKES	RADIOS	LAMPS	ROOFS	VALUE	
2	7	4	16	5	295,000	2	5	20	64 500	1	2	3	0	89,100	448,600

INCOME				FOOD SECURITY				HOUSEHOLD SIZE		
CROP SALES	ASSET SALES	OFF FARM INCOME	TOTAL INCOME	BOUGHT GRAIN	RATIONED FOOD	MEALS/DAY ADULT	MEALS/DAY CHILD	ADULTS	CHILD	60+
48 800	41 900	80 000	170 700	YES	NO	3	4	5	13	1

HOUSEHOLD INFORMATION CARDS — HOUSEHOLD NO. 1.55.1

ANIMALS						FARM TOOLS				CONSUMER GOODS					TOTAL ASSET VALUE
BULLS	COWS	GOATS	SHEEP	FOWLS	VALUE	PLOUGH	CUTLASS	HOES	VALUE	BIKES	RADIOS	LAMPS	ROOFS	VALUE	
0	0	3	0	5	13 500	0	3	8	6 300	1	0	5	0	21,500	41 300

INCOME				FOOD SECURITY				HOUSEHOLD SIZE		
CROP SALES	ASSET SALES	OFF FARM INCOME	TOTAL INCOME	BOUGHT GRAIN	RATIONED FOOD	MEALS/DAY ADULT	MEALS/DAY CHILD	ADULTS	CHILD	60+
10 000	9 600	0	19 600	YES	YES	2	3	3	5	0

HOUSEHOLD INFORMATION CARDS — HOUSEHOLD NO. 1.58.2

ANIMALS						FARM TOOLS				CONSUMER GOODS					TOTAL ASSET VALUE
BULLS	COWS	GOATS	SHEEP	FOWLS	VALUE	PLOUGH	CUTLASS	HOES	VALUE	BIKES	RADIOS	LAMPS	ROOFS	VALUE	
0	0	1	0	1	4 100	0	2	3	2 800	0	0	1	0	700	7 600

INCOME				FOOD SECURITY				HOUSEHOLD SIZE		
CROP SALES	ASSET SALES	OFF FARM INCOME	TOTAL INCOME	BOUGHT GRAIN	RATIONED FOOD	MEALS/DAY ADULT	MEALS/DAY CHILD	ADULTS	CHILD	60+
0	2 200	7 400	9 600	YES	YES	1	2	2	2	0

HOUSEHOLD INFORMATION CARDS

HOUSEHOLD NO. 1.15.1

ANIMALS						FARM TOOLS				CONSUMER GOODS					TOTAL ASSET VALUE
BULLS	COWS	GOATS	SHEEP	FOWLS	VALUE	PLOUGH	CUTLASS	HOES	VALUE	BIKES	RADIOS	LAMPS	ROOFS	VALUE	
9	12	25	2	46	826 100	2	16	29	75 400	2	0	19	10	119 300	1 100 800

INCOME				FOOD SECURITY				HOUSEHOLD SIZE		
CROP SALES	ASSET SALES	OFF FARM INCOME	TOTAL INCOME	BOUGHT GRAIN	RATIONED FOOD	MEALS/DAY		ADULTS	CHILD	60+
						ADULT	CHILD			
36 550	118 300	296 000	450 850	YES	NO	3	3	14	16	0

HOUSEHOLD INFORMATION CARDS

HOUSEHOLD NO. 1.16.1

ANIMALS						FARM TOOLS				CONSUMER GOODS					TOTAL ASSET VALUE
BULLS	COWS	GOATS	SHEEP	FOWLS	VALUE	PLOUGH	CUTLASS	HOES	VALUE	BIKES	RADIOS	LAMPS	ROOFS	VALUE	
0	1	0	2	0	26 000	1	1	7	29 700	1	0	0	0	18 000	73 700

INCOME				FOOD SECURITY				HOUSEHOLD SIZE		
CROP SALES	ASSET SALES	OFF FARM INCOME	TOTAL INCOME	BOUGHT GRAIN	RATIONED FOOD	MEALS/DAY		ADULTS	CHILD	60+
						ADULT	CHILD			
56 900	18 500	26 000	101 400	YES	YES	2	3	5	6	0

HOUSEHOLD INFORMATION CARDS

HOUSEHOLD NO. 1.17.1

ANIMALS						FARM TOOLS				CONSUMER GOODS					TOTAL ASSET VALUE
BULLS	COWS	GOATS	SHEEP	FOWLS	VALUE	PLOUGH	CUTLASS	HOES	VALUE	BIKES	RADIOS	LAMPS	ROOFS	VALUE	
2	0	3	3	6	113 100	1	3	6	30 100	1	0	0	0	18 000	161 200

INCOME				FOOD SECURITY				HOUSEHOLD SIZE		
CROP SALES	ASSET SALES	OFF FARM INCOME	TOTAL INCOME	BOUGHT GRAIN	RATIONED FOOD	MEALS/DAY		ADULTS	CHILD	60+
						ADULT	CHILD			
3 200	51 400	48 000	102 600	YES	YES	2	3	5	14	0

HOUSEHOLD INFORMATION CARDS

HOUSEHOLD NO. 1.23.1

ANIMALS						FARM TOOLS				CONSUMER GOODS					TOTAL ASSET VALUE
BULLS	COWS	GOATS	SHEEP	FOWLS	VALUE	PLOUGH	CUTLASS	HOES	VALUE	BIKES	RADIOS	LAMPS	ROOFS	VALUE	
1	0	2	5	11	73 600	1	3	6	30 100	1	0	3	0	20 100	123 800

INCOME				FOOD SECURITY				HOUSEHOLD SIZE		
CROP SALES	ASSET SALES	OFF FARM INCOME	TOTAL INCOME	BOUGHT GRAIN	RATIONED FOOD	MEALS/DAY		ADULTS	CHILD	60+
						ADULT	CHILD			
14 000	11 500	2 500	28 000	YES	YES	3	3	4	11	0

HOUSEHOLD INFORMATION CARDS — HOUSEHOLD NO. 1.64.1

ANIMALS						FARM TOOLS				CONSUMER GOODS					TOTAL ASSET VALUE
BULLS	COWS	GOATS	SHEEP	FOWLS	VALUE	PLOUGH	CUTLASS	HOES	VALUE	BIKES	RADIOS	LAMPS	ROOFS	VALUE	
0	0	0	0	2	1 200	0	3	7	5 700	0	0	0	0	0	6 900

INCOME				FOOD SECURITY				HOUSEHOLD SIZE		
CROP SALES	ASSET SALES	OFF FARM INCOME	TOTAL INCOME	BOUGHT GRAIN	RATIONED FOOD	MEALS/DAY		ADULTS	CHILD	60+
						ADULT	CHILD			
0	0	29 100	29 100	YES	YES	2	2	2	4	1

HOUSEHOLD INFORMATION CARDS — HOUSEHOLD NO. 2.02.1

ANIMALS						FARM TOOLS				CONSUMER GOODS					TOTAL ASSET VALUE
BULLS	COWS	GOATS	SHEEP	FOWLS	VALUE	PLOUGH	CUTLASS	HOES	VALUE	BIKES	RADIOS	LAMPS	ROOFS	VALUE	
1	1	2	4	5	87 000	0	6	13	10 800	2	1	8	5	128 600	226 400

INCOME				FOOD SECURITY				HOUSEHOLD SIZE		
CROP SALES	ASSET SALES	OFF FARM INCOME	TOTAL INCOME	BOUGHT GRAIN	RATIONED FOOD	MEALS/DAY		ADULTS	CHILD	60+
						ADULT	CHILD			
49 400	7 400	134 000	190 800	YES	NO	3	3	8	12	1

HOUSEHOLD INFORMATION CARDS — HOUSEHOLD NO. 2.05.1

ANIMALS						FARM TOOLS				CONSUMER GOODS					TOTAL ASSET VALUE
BULLS	COWS	GOATS	SHEEP	FOWLS	VALUE	PLOUGH	CUTLASS	HOES	VALUE	BIKES	RADIOS	LAMPS	ROOFS	VALUE	
2	4	4	7	25	220 000	1	3	10	32 500	2	2	10	2	97 000	349 500

INCOME				FOOD SECURITY				HOUSEHOLD SIZE		
CROP SALES	ASSET SALES	OFF FARM INCOME	TOTAL INCOME	BOUGHT GRAIN	RATIONED FOOD	MEALS/DAY		ADULTS	CHILD	60+
						ADULT	CHILD			
32 000	21 500	400 000	453 500	YES	N	3	3	9	12	1

HOUSEHOLD INFORMATION CARDS — HOUSEHOLD NO. 2.05.2

ANIMALS						FARM TOOLS				CONSUMER GOODS					TOTAL ASSET VALUE
BULLS	COWS	GOATS	SHEEP	FOWLS	VALUE	PLOUGH	CUTLASS	HOES	VALUE	BIKES	RADIOS	LAMPS	ROOFS	VALUE	
2	1	5	6	22	158 700	1	4	5	30 000	2	2	5	3	108 500	297 200

INCOME				FOOD SECURITY				HOUSEHOLD SIZE		
CROP SALES	ASSET SALES	OFF FARM INCOME	TOTAL INCOME	BOUGHT GRAIN	RATIONED FOOD	MEALS/DAY		ADULTS	CHILD	60+
						ADULT	CHILD			
160 900	0	153 600	314 500	NO	NO	3	3	8	13	1

HOUSEHOLD INFORMATION CARDS — HOUSEHOLD NO. 2.06.1

ANIMALS						FARM TOOLS				CONSUMER GOODS					TOTAL ASSET VALUE
BULLS	COWS	GOATS	SHEEP	FOWLS	VALUE	PLOUGH	CUTLASS	HOES	VALUE	BIKES	RADIOS	LAMPS	ROOFS	VALUE	
0	0	0	0	3	1 800	0	1	5	3 500	0	0	1	0	700	6 000

INCOME				FOOD SECURITY				HOUSEHOLD SIZE		
CROP SALES	ASSET SALES	OFF FARM INCOME	TOTAL INCOME	BOUGHT GRAIN	RATIONED FOOD	MEALS/DAY ADULT	MEALS/DAY CHILD	ADULTS	CHILD	60+
0	1 400	0	1 400	YES	YES	2	2	2	4	0

HOUSEHOLD INFORMATION CARDS — HOUSEHOLD NO. 2.07.1

ANIMALS						FARM TOOLS				CONSUMER GOODS					TOTAL ASSET VALUE
BULLS	COWS	GOATS	SHEEP	FOWLS	VALUE	PLOUGH	CUTLASS	HOES	VALUE	BIKES	RADIOS	LAMPS	ROOFS	VALUE	
2	10	6	5	9	331 400	1	3	12	33 700	3	3	17	9	236 900	602 000

INCOME				FOOD SECURITY				HOUSEHOLD SIZE		
CROP SALES	ASSET SALES	OFF FARM INCOME	TOTAL INCOME	BOUGHT GRAIN	RATIONED FOOD	MEALS/DAY ADULT	MEALS/DAY CHILD	ADULTS	CHILD	60+
13 200	8 000	672 000	765 200	YES	NO	3	3	16	11	1

HOUSEHOLD INFORMATION CARDS — HOUSEHOLD NO. 2.10.1

ANIMALS						FARM TOOLS				CONSUMER GOODS					TOTAL ASSET VALUE
BULLS	COWS	GOATS	SHEEP	FOWLS	VALUE	PLOUGH	CUTLASS	HOES	VALUE	BIKES	RADIOS	LAMPS	ROOFS	VALUE	
2	6	5	0	8	232 300	1	0	4	28 400	0	0	4	0	2 800	263 500

INCOME				FOOD SECURITY				HOUSEHOLD SIZE		
CROP SALES	ASSET SALES	OFF FARM INCOME	TOTAL INCOME	BOUGHT GRAIN	RATIONED FOOD	MEALS/DAY ADULT	MEALS/DAY CHILD	ADULTS	CHILD	60+
84 600	54 000	18 000	156 600	YES	NO	2	3	4	4	0

HOUSEHOLD INFORMATION CARDS — HOUSEHOLD NO. 2.12.1

ANIMALS						FARM TOOLS				CONSUMER GOODS					TOTAL ASSET VALUE
BULLS	COWS	GOATS	SHEEP	FOWLS	VALUE	PLOUGH	CUTLASS	HOES	VALUE	BIKES	RADIOS	LAMPS	ROOFS	VALUE	
0	0	0	0	0	0	0	1	4	2 900	0	0	0	0	0	2 900

INCOME				FOOD SECURITY				HOUSEHOLD SIZE		
CROP SALES	ASSET SALES	OFF FARM INCOME	TOTAL INCOME	BOUGHT GRAIN	RATIONED FOOD	MEALS/DAY ADULT	MEALS/DAY CHILD	ADULTS	CHILD	60+
0	0	18 800	18 800	YES	YES	2	2	2	4	1

HOUSEHOLD INFORMATION CARDS

HOUSEHOLD NO. 2.15.1

ANIMALS						FARM TOOLS				CONSUMER GOODS					TOTAL ASSET VALUE
BULLS	COWS	GOATS	SHEEP	FOWLS	VALUE	PLOUGH	CUTLASS	HOES	VALUE	BIKES	RADIOS	LAMPS	ROOFS	VALUE	
0	0	0	0	3	1 800	0	4	4	4 400	0	1	3	0	14 100	20 300

INCOME				FOOD SECURITY				HOUSEHOLD SIZE		
CROP SALES	ASSET SALES	OFF FARM INCOME	TOTAL INCOME	BOUGHT GRAIN	RATIONED FOOD	MEALS/DAY		ADULTS	CHILD	60+
						ADULT	CHILD			
23 200	800	10 500	34 500	YES	YES	2	3	4	4	1

HOUSEHOLD INFORMATION CARDS

HOUSEHOLD NO. 2.18.1

ANIMALS						FARM TOOLS				CONSUMER GOODS					TOTAL ASSET VALUE
BULLS	COWS	GOATS	SHEEP	FOWLS	VALUE	PLOUGH	CUTLASS	HOES	VALUE	BIKES	RADIOS	LAMPS	ROOFS	VALUE	
4	2	2	10	18	267 800	2	6	8	57 800	2	0	7	0	40 900	366 500

INCOME				FOOD SECURITY				HOUSEHOLD SIZE		
CROP SALES	ASSET SALES	OFF FARM INCOME	TOTAL INCOME	BOUGHT GRAIN	RATIONED FOOD	MEALS/DAY		ADULTS	CHILD	60+
						ADULT	CHILD			
44 900	74 600	54 000	173 500	YES	NO	2	3	6	3	2

HOUSEHOLD INFORMATION CARDS

HOUSEHOLD NO. 2.20.1

ANIMALS						FARM TOOLS				CONSUMER GOODS					TOTAL ASSET VALUE
BULLS	COWS	GOATS	SHEEP	FOWLS	VALUE	PLOUGH	CUTLASS	HOES	VALUE	BIKES	RADIOS	LAMPS	ROOFS	VALUE	
0	0	4	0	7	18 200	0	3	8	6 300	0	0	1	0	700	25 200

INCOME				FOOD SECURITY				HOUSEHOLD SIZE		
CROP SALES	ASSET SALES	OFF FARM INCOME	TOTAL INCOME	BOUGHT GRAIN	RATIONED FOOD	MEALS/DAY		ADULTS	CHILD	60+
						ADULT	CHILD			
17 300	8 300	1 500	27 100	YES	YES	1	2	1	3	0

HOUSEHOLD INFORMATION CARDS

HOUSEHOLD NO. 2.42.1

ANIMALS						FARM TOOLS				CONSUMER GOODS					TOTAL ASSET VALUE
BULLS	COWS	GOATS	SHEEP	FOWLS	VALUE	PLOUGH	CUTLASS	HOES	VALUE	BIKES	RADIOS	LAMPS	ROOFS	VALUE	
2	2	0	0	20	142 000	1	7	20	40 500	1	0	4	1	35 800	218 300

INCOME				FOOD SECURITY				HOUSEHOLD SIZE		
CROP SALES	ASSET SALES	OFF FARM INCOME	TOTAL INCOME	BOUGHT GRAIN	RATIONED FOOD	MEALS/DAY		ADULTS	CHILD	60+
						ADULT	CHILD			
62 300	25 000	44 100	131 400	YES	YES	3	3	4	8	0

HOUSEHOLD INFORMATION CARDS														HOUSEHOLD NO. 1.02.1
ANIMALS						FARM TOOLS				CONSUMER GOODS					TOTAL ASSET VALUE
BULLS	COWS	GOATS	SHEEP	FOWLS	VALUE	PLOUGH	CUTLASS	HOES	VALUE	BIKES	RADIOS	LAMPS	ROOFS	VALUE	
0	0	4	0	1	14 600	0	2	6	4 600	0	0	2	0	1 400	20 600

INCOME				FOOD SECURITY				HOUSEHOLD SIZE		
CROP SALES	ASSET SALES	OFF FARM INCOME	TOTAL INCOME	BOUGHT GRAIN	RATIONED FOOD	MEALS/DAY ADULT	MEALS/DAY CHILD	ADULTS	CHILD	60+
0	0	28 800	28 800	YES	YES	2	-	2	0	1

HOUSEHOLD INFORMATION CARDS														HOUSEHOLD NO. 1.04.1
ANIMALS						FARM TOOLS				CONSUMER GOODS					TOTAL ASSET VALUE
BULLS	COWS	GOATS	SHEEP	FOWLS	VALUE	PLOUGH	CUTLASS	HOES	VALUE	BIKES	RADIOS	LAMPS	ROOFS	VALUE	
2	5	3	3	21	222 100	1	2	8	30 800	1	1	5	5	108 500	361 400

INCOME				FOOD SECURITY				HOUSEHOLD SIZE		
CROP SALES	ASSET SALES	OFF FARM INCOME	TOTAL INCOME	BOUGHT GRAIN	RATIONED FOOD	MEALS/DAY ADULT	MEALS/DAY CHILD	ADULTS	CHILD	60+
121 600	112 000	176 000	409 600	NO	NO	3	3	4	9	0

HOUSEHOLD INFORMATION CARDS														HOUSEHOLD NO. 1.09.1
ANIMALS						FARM TOOLS				CONSUMER GOODS					TOTAL ASSET VALUE
BULLS	COWS	GOATS	SHEEP	FOWLS	VALUE	PLOUGH	CUTLASS	HOES	VALUE	BIKES	RADIOS	LAMPS	ROOFS	VALUE	
0	0	5	3	4	28 900	0	2	5	4 000	1	0	3	2	50 100	83 000

INCOME				FOOD SECURITY				HOUSEHOLD SIZE		
CROP SALES	ASSET SALES	OFF FARM INCOME	TOTAL INCOME	BOUGHT GRAIN	RATIONED FOOD	MEALS/DAY ADULT	MEALS/DAY CHILD	ADULTS	CHILD	60+
20 800	6 800	128 000	155 600	YES	YES	3	3	2	3	0

HOUSEHOLD INFORMATION CARDS														HOUSEHOLD NO. 1.10.1
ANIMALS						FARM TOOLS				CONSUMER GOODS					TOTAL ASSET VALUE
BULLS	COWS	GOATS	SHEEP	FOWLS	VALUE	PLOUGH	CUTLASS	HOES	VALUE	BIKES	RADIOS	LAMPS	ROOFS	VALUE	
3	6	4	10	30	317 000	2	15	25	72 500	1	1	7	6	124 900	514 400

INCOME				FOOD SECURITY				HOUSEHOLD SIZE		
CROP SALES	ASSET SALES	OFF FARM INCOME	TOTAL INCOME	BOUGHT GRAIN	RATIONED FOOD	MEALS/DAY ADULT	MEALS/DAY CHILD	ADULTS	CHILD	60+
101 800	51 200	434 000	587 000	YES	NO	2	3	9	13	0

HOUSEHOLD INFORMATION CARDS

HOUSEHOLD NO. 1.29.1.

ANIMALS						FARM TOOLS				CONSUMER GOODS					TOTAL ASSET VALUE
BULLS	COWS	GOATS	SHEEP	FOWLS	VALUE	PLOUGH	CUTLASS	HOES	VALUE	BIKES	RADIOS	LAMPS	ROOFS	VALUE	
1	0	5	3	8	76 300	1	0	5	28 000	1	1	4	3	77 800	182 100

INCOME				FOOD SECURITY				HOUSEHOLD SIZE		
CROP SALES	ASSET SALES	OFF FARM INCOME	TOTAL INCOME	BOUGHT GRAIN	RATIONED FOOD	MEALS/DAY		ADULTS	CHILD	60+
						ADULT	CHILD			
58 200	3 500	8 000	69 700	YES	YES	3	3	3	6	0

HOUSEHOLD INFORMATION CARDS

HOUSEHOLD NO. 1.31.1

ANIMALS						FARM TOOLS				CONSUMER GOODS					TOTAL ASSET VALUE
BULLS	COWS	GOATS	SHEEP	FOWLS	VALUE	PLOUGH	CUTLASS	HOES	VALUE	BIKES	RADIOS	LAMPS	ROOFS	VALUE	
2	0	5	0	12	114 700	1	7	13	36 300	1	1	5	2	63 500	214 500

INCOME				FOOD SECURITY				HOUSEHOLD SIZE		
CROP SALES	ASSET SALES	OFF FARM INCOME	TOTAL INCOME	BOUGHT GRAIN	RATIONED FOOD	MEALS/DAY		ADULTS	CHILD	60+
						ADULT	CHILD			
33 800	8 800	40 000	82 600	YES	NO	3	4	4	7	1

HOUSEHOLD INFORMATION CARDS

HOUSEHOLD NO. 1.33.1

ANIMALS						FARM TOOLS				CONSUMER GOODS					TOTAL ASSET VALUE
BULLS	COWS	GOATS	SHEEP	FOWLS	VALUE	PLOUGH	CUTLASS	HOES	VALUE	BIKES	RADIOS	LAMPS	ROOFS	VALUE	
2	0	3	7	20	133 500	1	9	12	36 700	1	2	6	4	106 200	276 400

INCOME				FOOD SECURITY				HOUSEHOLD SIZE		
CROP SALES	ASSET SALES	OFF FARM INCOME	TOTAL INCOME	BOUGHT GRAIN	RATIONED FOOD	MEALS/DAY		ADULTS	CHILD	60+
						ADULT	CHILD			
210 800	12.000	113 000	337 800	YES	NO	3	3	9	12	0

HOUSEHOLD INFORMATION CARDS

HOUSEHOLD NO. 1.34.1

ANIMALS						FARM TOOLS				CONSUMER GOODS					TOTAL ASSET VALUE
BULLS	COWS	GOATS	SHEEP	FOWLS	VALUE	PLOUGH	CUTLASS	HOES	VALUE	BIKES	RADIOS	LAMPS	ROOFS	VALUE	
2	2	0	3	1	139 600	1	1	4	27 900	0	0	4	0	2 800	170 300

INCOME				FOOD SECURITY				HOUSEHOLD SIZE		
CROP SALES	ASSET SALES	OFF FARM INCOME	TOTAL INCOME	BOUGHT GRAIN	RATIONED FOOD	MEALS/DAY		ADULTS	CHILD	60+
						ADULT	CHILD			
12 000	3 500	1 000	16 500	YES	YES	2	3	3	1	2

HOUSEHOLD INFORMATION CARDS — HOUSEHOLD NO. 2.44.1

ANIMALS						FARM TOOLS				CONSUMER GOODS					TOTAL ASSET VALUE
BULLS	COWS	GOATS	SHEEP	FOWLS	VALUE	PLOUGH	CUTLASS	HOES	VALUE	BIKES	RADIOS	LAMPS	ROOFS	VALUE	
2	0	0	3	4	101 400	1	3	6	30 100	2	1	4	0	50 800	182 300

INCOME				FOOD SECURITY				HOUSEHOLD SIZE		
CROP SALES	ASSET SALES	OFF FARM INCOME	TOTAL INCOME	BOUGHT GRAIN	RATIONED FOOD	MEALS/DAY		ADULTS	CHILD	60+
						ADULT	CHILD			
27 000	0	64 000	91 000	YES	YES	3	3	5	3	2

HOUSEHOLD INFORMATION CARDS — HOUSEHOLD NO. 2.58.1

ANIMALS						FARM TOOLS				CONSUMER GOODS					TOTAL ASSET VALUE
BULLS	COWS	GOATS	SHEEP	FOWLS	VALUE	PLOUGH	CUTLASS	HOES	VALUE	BIKES	RADIOS	LAMPS	ROOFS	VALUE	
2	0	4	2	24	124 400	0	4	12	9 200	0	0	6	4	64 200	197 800

INCOME				FOOD SECURITY				HOUSEHOLD SIZE		
CROP SALES	ASSET SALES	OFF FARM INCOME	TOTAL INCOME	BOUGHT GRAIN	RATIONED FOOD	MEALS/DAY		ADULTS	CHILD	60+
						ADULT	CHILD			
38 600	25 000	28 000	91 600	YES	YES	3	3	7	8	0

HOUSEHOLD INFORMATION CARDS — HOUSEHOLD NO. 2.59.1

ANIMALS						FARM TOOLS				CONSUMER GOODS					TOTAL ASSET VALUE
BULLS	COWS	GOATS	SHEEP	FOWLS	VALUE	PLOUGH	CUTLASS	HOES	VALUE	BIKES	RADIOS	LAMPS	ROOFS	VALUE	
4	10	3	8	14	422 900	1	1	14	33 900	1	2	7	1	316 900	773 700

INCOME				FOOD SECURITY				HOUSEHOLD SIZE		
CROP SALES	ASSET SALES	OFF FARM INCOME	TOTAL INCOME	BOUGHT GRAIN	RATIONED FOOD	MEALS/DAY		ADULTS	CHILD	60+
						ADULT	CHILD			
39 000	0	874 000	913 000	YES	NO	3	4	4	12	0

HOUSEHOLD INFORMATION CARDS — HOUSEHOLD NO. 2.62.1

ANIMALS						FARM TOOLS				CONSUMER GOODS					TOTAL ASSET VALUE
BULLS	COWS	GOATS	SHEEP	FOWLS	VALUE	PLOUGH	CUTLASS	HOES	VALUE	BIKES	RADIOS	LAMPS	ROOFS	VALUE	
1	2	2	2	6	101 600	0	3	5	4 500	1	0	3	0	20 100	126 200

INCOME				FOOD SECURITY				HOUSEHOLD SIZE		
CROP SALES	ASSET SALES	OFF FARM INCOME	TOTAL INCOME	BOUGHT GRAIN	RATIONED FOOD	MEALS/DAY		ADULTS	CHILD	60+
						ADULT	CHILD			
6 600	39 400	88 000	134 000	YES	YES	3	3	3	2	1

Why do we need to understand perceptions of risk?

Purpose

This activity demonstrates how perceptions of risk are related to socio-economic and cultural factors, as well as gender.

> **Note**
>
> The particular conditions under which people live impact on how they perceive risk. Development workers need to understand their own perceptions of risk. These may differ from those that inform decision-making at individual and community levels.

Procedure

This is a simulation in which a participant - playing the role of 'woman' - is confronted with a variety of options, all of which present additional pressures. Participants are asked to argue for or against options on the basis of differently perceived risk.

Time

- 1 hour

Materials

- role descriptions (see resources)

Process

Introduction

1. Introduce the activity by telling a story that illustrates how perceptions of risk vary greatly; give an example taken from a typical situation in daily life in order to illustrate different perceptions of risk.

> ***Sample story***
>
> This is a sample story; try to think of your own or ask one of the participants to offer an example.
>
> *Mrs. Kahuma knows that it is illegal to make charcoal: the government has said that the charcoal-making uses up too much wood, and this poses a risk to the environment. A little while ago an official caught her neighbour when she was trying to sell charcoal, and she had to pay a heavy fine or go to prison. But if she could sell her pile of charcoal today, she could go to the shop and buy some candles and tea...*

2. Ask participants to list the risk factors that women are exposed to:

   — Firstly, in their various reproductive, productive and community management roles, and

   — Secondly, due to cultural, religious or other factors;

   Record them on flipchart.

3. Point out that the following simulation asks participants to explore some of these risks.

Participant Action

1. Explain that this simulation requires 11 people:

   5 to act as observers; 5 to act as 'options'; 1 to act as 'woman'.

   Ask for volunteers and allocate roles; ensure that the actress/actor of the part 'woman' is assertive and able to interrogate her 'proposers'.

2. Hand each of the observers, 'options' and the 'woman' their role descriptions (see resources), and ask them to read through it.

3. Ask 'options' to stand in a circle; ask each observer to positon her/himself behind one of the 'options'. Ask 'woman' to enter the circle.

4. Outline the process of the simulation:

   There are three 'rounds' to this simulation.

   Round 1: 'woman' turns to each of the 'options' and they make her an 'offer'.

   Round 2: she attempts to find out more about the 'offer', and they give her information. Bit by bit, she enters into discussion with them, trying to assess the risk involved.

   Round 3: she is left to make a decision. Participants will be asked to advise her.

   The observers will record the process and at the end, feed back some of the observations and information gathered.

5. Begin the process:

   — Introduce 'woman' by reading out her role description:

   This is Natsai; she lives in a small house in a village some 120 Km from the nearest larger town. Natsai has attended school for four years. She had her first child when she was 17; she now has four children; the youngest two are under five. Her husband left the village like many men, in search of work; his remittances stopped over two years ago. Natsai has a small plot of land about 20 minutes walk away from the village.

   Her area has been hit repeatedly by severe droughts and the land is badly degraded and eroded. Flashfloods occur in the rainy season. It is unclear how many villagers are HIV positive, but many young adults suffer from chronic chest infections.

   — Outline the present situation:

   It is winter. For weeks Natsai has not been able to feed her children anything but porridge. Her food reserves are depleted.

   Unable to continue life in this manner, Natsai tries to make a decision as to what to do.

   In the course of trying to make a decision Natsai is propositioned by various 'options'.

6. Begin round 1: Spin 'Natsai' around randomly, until she faces one of the 'options'. Allow for a brief exchange, then spin her again, until she faces another 'option'. Repeat this process until she has spoken to each of the 'options'.

7. Round 2: continue to spin 'Natsai'around, stopping for interrogations with various 'options', and allowing sufficient time for exchanges of information. Move 'Natsai' whenever the 'option' has run out of ideas. Continue this process until the descriptions of risks have been exhausted.

Review and Discussion

1. Ask observers to report what they saw / heard:
   - ? What did they see happening?
   - ? How did it happen, and why?
   - ? What were the options presented?

   Record the options in columns, on flipchart and write up the key risks related to each.

2. In plenary, discuss the options:
   - ? What are the advantages / disadvantages of each?
   - ? What are the risks involved? Identify the elements at risk;
   - ? Weigh up the risks in terms of considerations such as long-term / short-term risk; risk at personal, household, community level;

3. Ask participants to reflect **individually** on 'Natsai's' options. Ask them to reach a decision on what advice they would give her.
   (Allow 5 minutes for this)

Participant Action 2

1. Request participants to stand in a circle; ask 'Natsai' to stand in the middle.

2. Explain that 'Natsai' will turn to each participant and ask him or her for advice on which option to choose. Ask participants to explain the reasons for their decision and inform them that they may remain silent, if they so choose.

3. Monitor the process and record key advice on flipchart.

Review and Discussion

1. Discuss how the different advice given relates to different perceptions of risk.

2. De-brief participants; thank them for taking the risk of participating in this simulation.

3. Initiate a plenary discussion on the process. Ask questions such as the following:
   - ? What did you learn about risk?
   - ? What did you learn about perceived risk?
   - ? What did you learn about options open to the poor?
   - ? What did you learn about options open to women?

4. Ask a participant to sum up what s/he learnt from this activity.

Resources

ROLE PLAY

(you may want to change the role descriptions, depending on the context within which you are working):

Role of Woman

Your name is Natsai; you live in a small house in a village some 120 km from the nearest larger town. You have attended school for four years. You had her first child when you were 17; your now have four children; the youngest two are under five. Your husband left the village like many men, in search of work; his remittances stopped over two years ago. You have a small plot of land about 20 minutes walk away from the village.

Your area has been hit repeatedly by severe droughts and the land is badly degraded and eroded. Flashfloods occur in the rainy season. It is unclear how many villagers are HIV positive, but many people suffer from chronic chest infections.

Option: Migration

Present 'migration' as an exciting option; outline all the possibilities such as:

— there may be job opportunities in town; migrating to the urban area might bring financial independence with it.
— Urban life is exciting;

What are the possible risks?

Option: Prostitution

Present 'prostitution' as a viable alternative; outline the advantages, such as

— escape from the hard rural life
— financial independence

What are the possible risks?

Option: Take on handicraft

Present 'home-crafts' as a way out; outline the positive spin-off effects such as

— financial independence
— you can stay in your secure home-environment

What are the possible risks?

Option: Continue subsistence living

Present continued subsistence living as the only realistic option; outline the advantages, such as

— you have been able to survive in the past
— at least you know the risks involved

What are the possible risks?

Section 3:

Emergency Preparedness and Risk Reduction

Section 3 is made up of twelve participatory learning activities that further improve participants' ability to incorporate risk reduction into emergency response plans. The activities are a mixture of analytical thinking and strategic planning exercises; participants practice how to identify appropriate strategies for reducing vulnerability, and how to examine proposed action plans in terms of gender considerations. This section aims at developing participants' planning skills by asking them to consider the impact of various hazards on elements at risk.

Do you have what it takes to work in risk reduction?

Purpose

- Working as a group, participants attempt to juggle a great number of shoes, for as long as possible. In terms of risk reduction: participants practise working collectively and cooperatively.

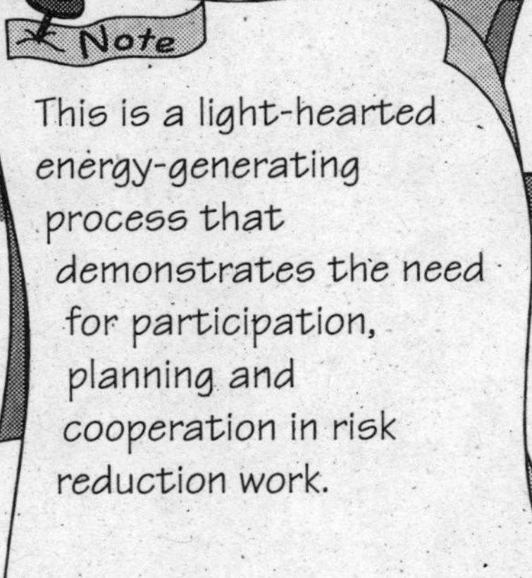

Note
This is a light-hearted energy-generating process that demonstrates the need for participation, planning and cooperation in risk reduction work.

Procedure

In this activity participants use their shoes as 'juggling balls'. This is followed by a discussion on risks and risk reduction.

Time

- ½ hour

Process

Introduction

1. Introduce the activity saying this is a 'Risk Reduction Juggle'. Request participants to take a small risk.

2. Ask participants to stand in a circle. Join in the circle.

3. Explain the following: in this activity we will juggle as many shoes as possible. To be able to do this, we have to plan and establish a definite order of throwing and catching, and then stick to that order all the way through.

Participant Action

1. Ask participants to take off one of their shoes, and you do likewise. Ask them to put the shoes on the ground, by their sides.

   (You may want to ask them to check that there is nothing stuck to the soles of their shoes.)

2. Holding your shoe in one hand explain the following:

   I will throw my shoe to one of you, across the room from me. To alert her/him, I will say her/his name. S/he will then throw the shoe to another participant across the room from her/him, again saying her/his name, and so on, until we have all had a turn at catching and throwing, and the shoe is back with me.

   It is important that you remember who to throw to, and who to catch from, and that you always follow the same order, throughout the juggle.

3. Check for understanding, and throw the shoe across the room, to another participant.

4. Watch carefully to check that participants follow your instructions. Ensure that no-one is left out or receives the shoe twice. When the shoe is back with you discuss what happened.

5. Repeat the process of throwing and catching until participants have established a clear order.

6. Explain that from now on, in each round participants will add another shoe: first the second participant, in the next round the third, and so on. Point out that many shoes juggled at the same time requires alertness.

7. Begin by throwing your shoe; prompt the next player to throw her/his shoe after s/he has passed on your shoe. Continue until many shoes are being thrown, or juggled, at the same time.

8. The process ends when the juggle collapses: this usually happens as soon as the pattern is broken and participants can no longer keep up with catching and throwing.

9. Allow participants to catch their breath, collect and put on their shoes and sit down.

Review and Discussion

1. Encourage participants to describe what happened; the juggle usually generates a lot of laughter and participants enjoy talking about the process.

2. Ask questions such as
   - ? What were the risks involved for each participant?
   - ? What was the risk involved in terms of the purpose of the activity: to juggle as many shoes as possible?

3. Ask participants to identify and list the attitudes and skills that were needed from each participant. Write them up on the flipchart.

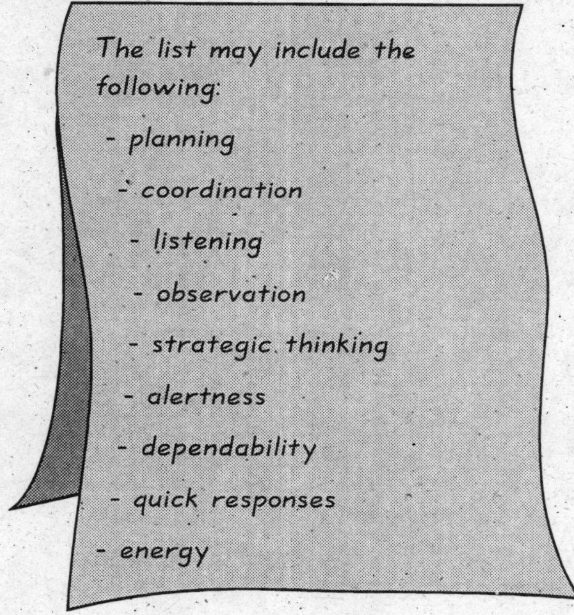

The list may include the following:
- planning
- coordination
- listening
- observation
- strategic thinking
- alertness
- dependability
- quick responses
- energy

4. Review the list: which of these attitudes and skills are typical of the qualities needed for community workers engaged in risk reduction? How are the qualities important? Who needs them?

5. Review the process of juggling: What made the juggle work, what made it collapse?
   — Discuss key issues such as participation, collective action, planning, cooperation, consultation etc.

6. Sum up by asking participants to make a brief statement about what they learnt from the 'risk reduction juggle'.

Notes

How do we plan for emergency preparedness and response?

Purpose

This series of activities strengthens participants' capacity to develop emergency preparedness and response plans specific to the needs of particular communities.

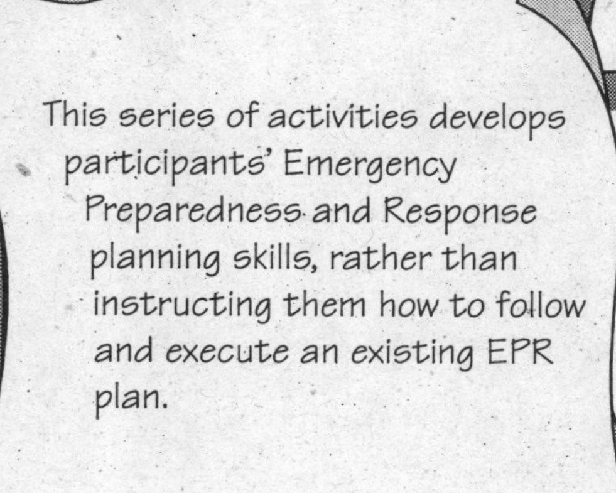

Note
This series of activities develops participants' Emergency Preparedness and Response planning skills, rather than instructing them how to follow and execute an existing EPR plan.

Procedure

The first activity is an information-giving session; this can be done either through a resource person / EPR expert, or by way of reading and discussion. The case study asks participants to apply the information gathered to a specific scenario. The fish-bowl discussion illustrates the need for coordinated planning and demonstrates the difference between preparedness and emergency planning.

Time

- approximately 5 hours

Materials

- reading material: (see resources)
  - EPR planning guidelines
  - role play and instructions
  - case story of a flood
  - technical information on floods

Process

Introduction to the session

1. Outline the purpose and procedure of the activity.

2. Point out that there is no one 'master-plan' for emergency preparedness and response and that plans are best developed in consultation and cooperation with the communities concerned.

   Remind participants that the information offered should serve as a guideline rather than a blueprint for EPR planning in the field.

3. Give an interactive input on EPR planning: list and explain the key questions that need to be asked when developing an EPR plan. (see resources)

4. Hand out reading materials (see resources)

   (a) guidelines for EPR planning

   (b) 'The 1987 Floods' - an eye-witness account of a flood experience

   (c) technical information on floods

Participant Action 1: reading and discussion

1. Ask participants to read their materials.
   Allow approximately 45 minutes for this.

Review and Discussion

1. Facilitate a brief plenary discussion on the reading:

   — Encourage questions of clarification.

   — Ensure understanding and common usage of specific disaster terms, such as hazard, risk, vulnerability.

   Allow approximately 15 minutes for this.

Participant Action 2: case study

1. Divide participants into three working groups representing NGOs and agencies working in the following areas:

   (i) health

   (ii) relief

   (iii) development, with a specific focus on income generating projects.

2. Set up the case study activity:

   — Hand each group copies of the case scenario and instructions.
   — Briefly read through the instructions and check for understanding.
   — Remind participants that they have 60 minutes to complete the task.

3. Manage and monitor the process by going from group to group, checking on progress and assisting if necessary.

Review and Discussion

1. Ask groups to reconvene in plenary, and facilitate report-backs. Allow approximately 10 minutes for each group's report.

2. Initiate and facilitate a plenary discussion. Ask questions such as the following:

   ? What information did you need? What sources of information did you draw on? How did you go about conducting the assessment?

   ? What were the problems you encountered when planning? How did you go about working together?

   ? What planning tools did you use? What resources did you draw on?

   ? Were there any overlaps between the EPR outlines presented? What were the similarities, what were areas that might lead to contestation?

3. Ask participants to apply the case study task to their work in the field: what were the problems and frustrations they encountered in the planning task, and how do they relate to real planning workshops?

4. Suggest an energiser / stretch or break.

Participant Action 3: No pressure fishbowl[1]

1. Introduce the next process.
   Explain that cooperation and coordination in emergency situations is crucial. Plans from different organisations / agencies should be coordinated in order to avoid duplication of actions or even conflict.

   Remind participants of the purpose of this activity:

   You will be asked to coordinate the plans developed in the previous activity. This will allow you to practise a number of skills, such as meeting skills, cooperative group work skills and planning skills. It will also serve as a means to demonstrate the different dynamics that arise in preparedness and emergency meetings.

2. Request each of the case study groups to appoint or elect a representative. Ask all other participants to act as observers.

3. Request participants to move into a fishbowl arrangement: The three group representatives will sit in the middle with all others as observers in a circle around them.

4. Give the following instructions:

   (a) to group representatives:

   You are asked to coordinate your EPR outlines and work towards one consolidated EPR plan.

   (b) to observers:

   focus on process and content of the task group with two main guiding questions in mind:

   (i) What did the group do to coordinate successfully?

   (ii) What aspects of EPR planning were overlooked / left out?

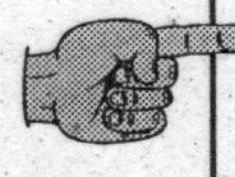

This planning meeting acted out by group representatives is likely to be fairly unpressurised and relaxed (no time limit was given!) and concrete outcomes are unexpected.

5. Allow approx. 15-20 minutes for the planning discussion.

6. Interrupt the process and thank the participants.

Review and Discussion

1. Facilitate a report from the observers. Point out that group representatives will be given a chance to respond to observations only after observers have completed their reports.

2. Record the key observations on crucial aspects of the coordination of plans on flipchart.

3. Ask participants to discuss what kind of a planning meeting this had been.

4. Point out that emergency preparedness meetings conducted when there is no imminent crisis are likely to be less pressurised and task-oriented than EPR planning meetings in times of an emergency.

Participant Action 4: High pressure fishbowl

Introduction

1. Request three new representatives (one from each group); ask them to sit in the 'fishbowl' and give the following instruction:

Coordinate the EPR outlines from the three planning groups.
You have 10 minutes to develop an action plan.

Participant Action

1. While the planning meeting happens, put pressure on the group and ensure they take the task seriously and in the spirit of an emergency situation.
   The following are some suggestions for creating an atmosphere of pressure:
   - give the instruction as an 'order' by speaking in a clipped way;
   - write up the 10 minutes as figures 1-10 and cross out numbers as the minutes pass;
   - draw a clock and show the passing of time by moving the 'finger'-line;
   - walk around the group, impatiently;
   - terminate the process when the time has elapsed.

Review and Discussion

1. Discuss what happened; record important observations on flipchart.

2. Identify the difference between the two planning meetings.

3. Ask participants to describe what a planning meeting might look like once the emergency has already started?

   — Point out that in times of real emergencies action replaces planning meetings; this is why preparedness plans are so crucial.

4. Ask participants to list some of the questions they would ask after an immediate emergency has passed; Point out that post-emergency meetings would serve as opportunities for evaluating EPR plans.

Questions might include:

What was the emergency response action supposed to be?:

What did we actually do?:

What was the difference?:

Was there duplication?:

What were the gaps in planning?

Session Review

As a way of summing up the series of activities: ask participants to take turns and one by one make brief statements about what they have learnt from these activities.

1. Thank you to Juan Saenz, International Federation of Red Cross and Red Crescent Societies, East Africa, for his ideas and facilitation of a session in 1995.

Resources

GUIDELINES

Emergency preparedness and response planning

1. **What do you know about the hazard and community at risk?**

   Have you (or has someone else) collected information about the past impact of this hazard on the community?

   Have you (or has someone else) carried out a community risk assessment (e.g. assessed the hazard and the community's vulnerabilities and capacities) ?

2. **In which sectors is your agency likely to respond?**
   (How does your agency's orientation in general relate to this type of threat? What types of services and support could it dependably offer?).

   What geographic area does your agency cover?

   Which target groups do you assist and how many?

   What types of assistance do you generally offer?

   How do your agency's goals and action plans fit and match with those of your partner organisations?

3. **Who will be in charge, and who will coordinate?**
   (Where will your focal response points be?).

   Who is in charge at local level/intermediate level/head office?

   Who is the focal point (give and circulate name and contact information within your agency and to other partners).

   Who are the team members (at each operational level)? What are their names and responsibilities?

   Is there a lead agency? Who? What are their responsibilities to your agency - and your responsibilities to them?

4. **How do you guarantee supplies, support and communication to the worst affected area?**

   What material supplies will you need?

   Where can you find them?

   How will you store and transport them?

   Can you find other NGOs/private companies to do transport for you?

What types of transport would you need to guarantee access during the worst of the emergency?

Will certain areas be cut off? If so, do you need to preposition supplies in advance? Who will watch over them?

How will you work with the community? (What community structures are there and who are the key people?)

How do you ensure communication between the local and other levels (assume there are no telephones)? What informal communication systems already exist for disseminating information?

Can you guarantee access to electricity? If not, what back-up supplies will you need to work?

5. **How and when do you activate a response? Who does it?**

   How do you track the warning information? Who is responsible?

   How do you track the impact information? Who is responsible?

   How do you disseminate the information within your agency? Who is responsible?

6. **What specific actions are you planning?**
   Define your actions for:

   — evacuation

   — emergency shelter

   — food and water supplies

   — medical care

   — public health services

   — transport

   — etc.

7. **What resources do you need to respond?**

   What do you already have and in what quantities?

   Can you shift or reallocate personnel, vehicles, funding supplies from other programmes?

   What else might you need?

   Can you link with other agencies to get these items and supplies?

   Who are other possible agencies and suppliers you can draw on locally?

8. **Who will be the key people and what are their responsibilities?**

- Make lists of people and their contact addresses and numbers for your agency.
- Make lists of people and their contact addresses and numbers for other agencies. For example:

Emergency Operation Centre:		Indundassion Settlements	
Field Officer in Charge:		Margaret Puddle	
Telephone: 73465		**Fax:** 73466	
TEAM MEMBERS			
Responsible for	**Name**	**Address**	**Telephone**
Registration	Mary Dlamini	11 Phillips Ave Belgravia Harare	76614

Emergency Operation Centre:		Indundassion Settlements	
Field Officer in Charge:		Margaret Puddle	
Telephone: 73465		**Fax:** 73466	
Other Agencies in		Indundassion	
Name	**Officer in Charge**	**Address**	**Telephone**
Food for the World			
Christian Medical Committee			
Boreholes International			
Health Department			

ROLE PLAY

The Flood of the Manzi River

Briefing for members of HEALTH NGOs

Background

A sprawling informal settlement is growing on the north-east side of a large city. Originally, people migrated to the site as a result of a devastating flood in the North of the country, and due to rising unemployment in the rural areas. Many had fled from political fighting between the country's two main parties. By now there are an estimated 60000 people living on the steep slopes of the Manzi river, which is reduced to a trickle during the dry season.

Large numbers of fragile structures have been erected on the soft soil; the ground is severely eroded. Most of the vegetation has been cut down for fuel. There is no infrastructure apart from communal water taps along the roadside. There is no sanitation. The nearest health clinic is 35 km away. There are no schools in this area. Adjacent to the settlement is a middle-class suburb.

Recently, there have been heavy rains and more rain is forecast. In the previous season the river flooded and 12 people drowned; thousands lost their dwellings. Cases of dysentery have been recorded; in the past months there has been an increase in the reported cases of measles due to poor to non-existent immunisation coverage.

Your task

You are a representative of an NGO that is concerned with issues of health, working in the Manzi settlement.

You attend a workshop of Health NGOs called to draft an Emergency Preparedness and Response (EPR) plan for the Manzi settlement.

PLEASE PREPARE AN OUTLINE OF AN EPR PLAN BASED ON THE SCENARIO GIVEN.

— You have 60 minutes to complete this task.

— Please prepare to do a 10 minutes presentation of your EPR outline to the rest of the group.

Briefing for members of RELIEF AGENCY

Background

A sprawling informal settlement is growing on the north-east side of a large city. Originally, people migrated to the site as a result of a devastating flood in the North of the country, and due to rising unemployment in the rural areas. Many had fled from political fighting between the country's two main parties. By now there are an estimated 60000 people living on the steep slopes of the Manzi river, which is reduced to a trickle during the dry season.

Large numbers of fragile structures have been erected on the soft soil; the ground is severely eroded. Most of the vegetation has been cut down for fuel. There is no infrastructure apart from communal water taps along the roadside. There is no sanitation. The nearest health clinic is 35 km away. There are no schools in this area. Adjacent to the settlement is a middle-class suburb.

Recently, there have been heavy rains and more rain is forecast. In the previous season the river flooded and 12 people drowned; thousands lost their dwellings. Cases of dysentery have been recorded; in the past months there has been an increase in the reported cases of measles due to poor to non-existent immunisation coverage.

Your task

You are a representative of a relief agency concerned with flood emergencies in the Manzi settlement.

You attend a workshop of relief agencies called to draft an Emergency Preparedness and Response (EPR) plan for the Manzi settlement.

PLEASE PREPARE AN OUTLINE OF AN EPR PLAN BASED ON THE SCENARIO GIVEN.

- You have 60 minutes to complete this task.
- Please prepare to do a 10 minutes presentation of your EPR outline to the rest of the group.

Briefing for members of DEVELOPMENT NGOs

Background

A sprawling informal settlement is growing on the north-east side of a large city. Originally, people migrated to the site as a result of a devastating flood in the North of the country, and due to rising unemployment in the rural areas. Many had fled from political fighting between the country's two main parties. By now there are an estimated 60000 people living on the steep slopes of the Manzi river, which is reduced to a trickle during the dry season.

Large numbers of fragile structures have been erected on the soft soil; the ground is severely eroded. Most of the vegetation has been cut down for fuel. There is no infrastructure apart from communal water taps along the roadside. There is no sanitation. The nearest health clinic is 35 km away. There are no schools in this area. Adjacent to the settlement is a middle-class suburb.

Recently, there have been heavy rains and more rain is forecast. In the previous season the river flooded and 12 people drowned; thousands lost their dwellings. Cases of dysentery have been recorded; in the past months there has been an increase in the reported cases of measles due to poor to non-existent immunisation coverage.

Your task

You are a representative of a development NGO working specifically with setting up and organising income generating projects in the Manzi settlement.

You attend a workshop of other development NGOs called to draft an Emergency Preparedness and Response (EPR) plan for the Manzi settlement.

PLEASE PREPARE AN OUTLINE OF AN EPR PLAN BASED ON THE SCENARIO GIVEN.

- You have 60 minutes to complete this task.
- Please prepare to do a 10 minutes presentation of your EPR outline to the rest of the group.

READING

The 1987 Floods[1]

In 1984 I lived at Lindelani, the shack area near Ntuzuma township. There were many of us there living in imijondolo. We were waiting for land where we could build proper houses and not be chased away. We were still waiting when the 1987 floods came.

When we woke up on Saturday 23 September 1987 it was raining softly. We did not know that the rain would carry on for four days and wash away our houses and families. At first the rain was soft but as the day went by the rain became very hard. It rained all day and all the time; it did not stop. We stayed inside all day and heard the rain outside. When we went to bed it was still raining and the bed felt damp.

I woke up in the middle of the night. It was raining even harder and I could feel that something terrible was going to happen. I was cold. I lit a candle. The roof was leaking and I looked for more pots to catch the drops. I could hear the water rushing by outside and I was afraid.

The next morning my younger sister, Thandekile, came to visit me. She said, "Hawu! The roads look like rivers. They are full of mud and water. All my clothes are wet!"

We made tea and listened to the rain. It rained all day long without a break. We tried to keep the water out of the house. My next door neighbours had a mud house and it was being worn away by the water which flowed around it the whole day. At about eight o'clock that night their house fell down. We heard their screams but we could not help them. The path between our houses had become a river we could not cross.

In the morning we looked out. We were shocked to see how these neighbours had been washed out. They lay dead in the mud. Some were without clothes, their naked bodies covered with mud. While we stood there in horror we saw something moving in the mud. We ran to that place and started digging. We found a child trapped between the corrugated iron and the wall of the house. We didn't think she would live. We worked for hours in the mud and the rain and finally we got her out alive. Six people were killed in that disaster, the parents and four of their children.

It rained again right through the night and in the morning it had not stopped. This was the third day of staying in the house and we needed some food. I askedThandekile to go to the nearest shop. She did not want to go alone. "Please come with me, Thembakazi," she begged. "I am so afraid of these roads which are rivers."

I agreed to go with her. I remember we left the house at about 10 o'clock. It was grey outside and still raining.

We had walked only a short way when we heard a loud BANG that it hurt our ears.

We looked towards the graveyard where the noise had come from. Filthy black water burst out of the graves in the cemetery and went roaring down the road. It looked just like boiling water. In no time the road turned into a swollen river. Then I saw coffins being carried by the river, and corpses in the foul smelling water.

I was shocked . I saw corpses being rolled along in the river. The bodies appeared and disappeared under the water while I watched. It was horrible.

A man who was walking on the other side of the river tried to come across. He was knocked off his feet by the force of the water and washed away by the roaring river. He saved his own life by catching hold of a tree. He grabbed a branch and held on for his life.

We saw that we would not be able to get to the shop. We decided to go back home.

The roads we used to walk on were just rivers. All the houses we saw were damaged. Some people were inside their shacks trying to keep the water out. Some people were outside in the rain trying to keep their shacks from falling down. There were no dry, safe places for shack dwellers that day.

When we got back to our house the place was flooded. Everything was under water or washed away. I saw that I had nothing left in the world. No clothes, no shoes, no underwear, no blankets, no furniture, not even a stove. Everything was under water or carried away in the mud. I had only the clothes I was wearing and a large towel that I had wrapped around my waist. I have never felt so lost.

That night we looked for a place to sleep. We went with other people to a Putco bus. Everyone in the bus was crying. Some were crying for their children, some were crying for their relatives and some were crying for their neighbours. We were cold and wet and hungry. We had nothing. Luckily some strangers came and brought us food. Then we slept.

It rained again the whole of the next day. That night we did not sleep in the bus again. By that time many people had heard about how shack dwellers were suffering. Some people came to try and help me. We were taken to a church hall where we were warmly welcomed. We got fresh food, hot water and clothes. We were warm and dry. We ate and we slept. We stayed there while it went on raining.

On Wednesday 27 September 1987 it began to stop raining. The next day we were taken back to Lindelani. When we got to Lindelani we were confused. There was not so much water now. The river was small. There was not so much noise now. The river was quiet. But we could not see our own houses. We did not know what to do. There were just bits of broken houses everywhere. The mud was full of planks and trees and rubbish. We started to search in the mud. We hoped that we would find some of our belongings.

I found a big plastic bag. When I opened it I saw something I could not believe. It was the pieces of a man. The bag had been buried but it was brought to the surface by the floods. The man's body had been chopped into small pieces. I thought he must be a victim of the violence that was happening at that time. When we walked on we found two more bodies. One was a man and the other was a woman. The woman's head was cut off and she had only one arm. The man's leg and his private parts were cut off. I thought this must be the place of the revenge killings.

The whole place was a mess. There was a bad smell all over Lindelani. This was because there were so many corpses and bones lying in the mud. We saw dogs feeding on a human body. It became part of our daily lives to see parts of bodies lying on the road. It was horrible.

On the third day after the rain stopped a large tent was erected by the Red Cross. Some people went there. At the same time some donations from the Flood Relief Fund were given to us.

After many months some land surveyors arrived at Lindelani. They called us one by one. They allocated plots to families and gave us tents to put up. It took a long time for each family to get a plot and a tent. Even then we were told that the a tents and the plots which we had been given were temporary. We waited until 1988 before we allocated permanent plots. We are still waiting for proper houses. Every year when it rains we hope the rain will not bring floods again.

1. Mnisi, T. (1991) *The 1987 Floods.* New Readers Project, Centre for Adult Education, University of Natal: Durban.

Technical information on floods[1]

FLOODS

This chapter of the module aims to improve your understanding of:

- *the causes of floods and factors which intensify their effects*
- *impacts of floods on human settlements*
- *flood control, prevention and preparedness measures*
- *flood forecasting and warning systems*

Introduction

Throughout history people have been attracted to the fertile lands of the floodplains where their lives have been made easier by virtue of close proximity to sources of food and water. Ironically, the same river or stream that provides sustenance to the surrounding population, also renders these populations vulnerable to disaster by periodic flooding. Floods can arise from abnormally heavy precipitation, dam failures, rapid snow melts, river blockages or even burst water mains. Flood disasters are second only to droughts in the total number of people affected world wide.

FLOOD HAZARD DATA SHEET

Number killed by declared flood disasters, 1980-89: 16,108
Number affected: 279,330,901 *(OFDA, 1990)*

Selected severe flood disasters

Year	Location	Deaths	Losses in US$ million
1966	Italy	113	1,300
1983	Peru, Ecuador	500	700
1983	Spain	42	1,250
1986	China	260	1,210
1987	USSR	110	550
1987	Switzerland	0	700
1987	Bangladesh	1,600	1,300
1988	Thailand	371	300
1988	Bangladesh	3,000	1,200
1990	Tunisia	25	211
1991	China	2,295	12,500

Source: Nature and Resources, Vol. 27, No.1, 1991

Causal phenomena

Types of floods

Flash floods — These are usually defined as floods which occur within six hours of the beginning of heavy rainfall, and are usually associated with towering cumulus clouds, severe thunderstorms, tropical cyclones or during the passage of cold weather fronts. This type of flooding requires rapid localized warnings and immediate response by affected communities if damage is to be mitigated. Flash floods are normally a result of runoff from a torrential downpour, particularly if the catchment slope is unable to absorb and hold a significant part of the water. Other causes of flash floods include dam failure or sudden breakup of ice jams or other river obstructions. Flash floods are potential threats particularly where the terrain is steep, surface runoff is high, water flows through narrow canyons and where severe rainstorms are likely.

Truck carried away by flooding

Mass Media Production Centre, Manila, *UNDRO News*, Sep/Oct 1984

River floods — River floods are usually caused by precipitation over large catchment areas or by melting of the winter's accumulation of snow or sometimes by both. The floods take place in river systems with tributaries that may drain large geographic areas and encompass many independent river basins. In contrast to flash floods, river floods normally build up slowly, are often seasonal and may continue for days or weeks. Factors governing the amount of flooding include ground conditions (the amount of moisture in the soil, vegetation cover, depth of snow, cover by impervious urban surfaces such as concrete) and size of the catchment basin. In some larger semi-arid countries, such as Australia, inland flooding of dry or stagnant rivers may occur many weeks after the onset of heavy coastal monsoon or cyclonic rain has directed river flows many hundreds of km inland, and in the complete absence of any sign of disturbed weather. Historical records of flooding of towns on the main river flood plains prove

Figure 2.2.1

Flooding and its causes

Natural Hazards, Disaster Management Center, 1989.

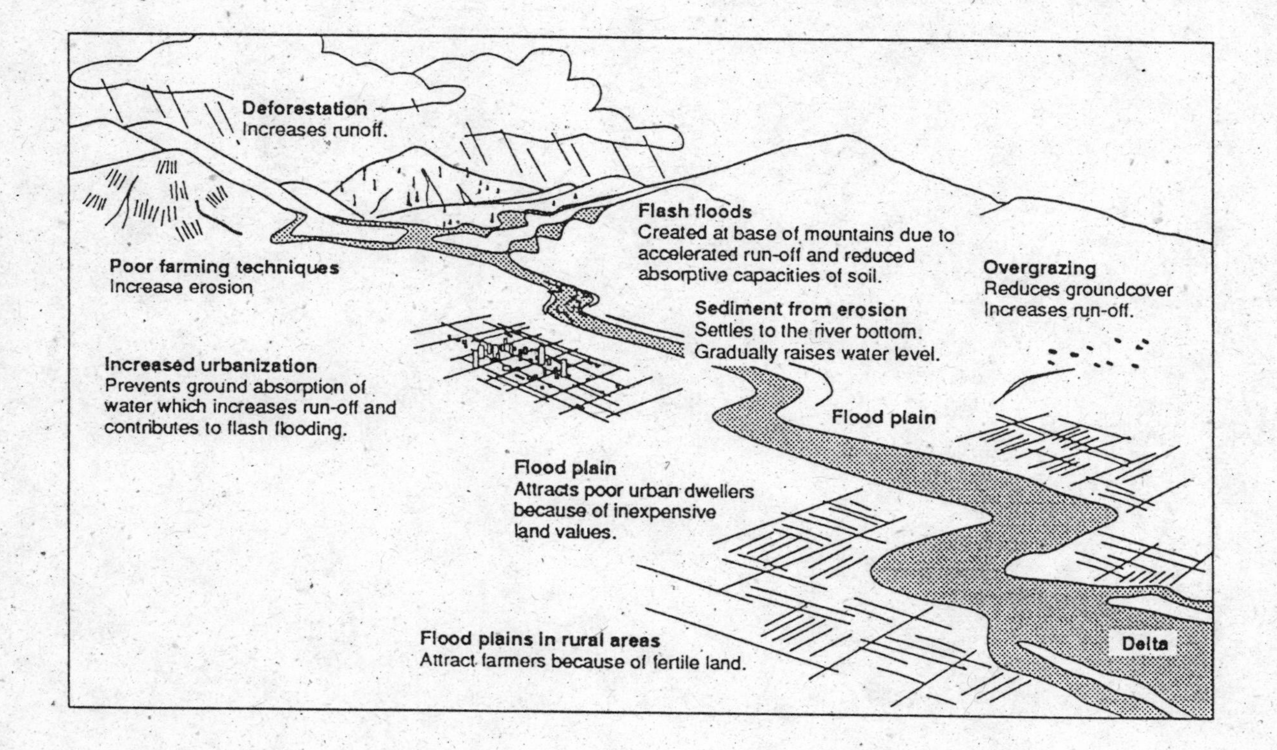
Deforestation
Increases runoff.
Poor farming techniques
Increase erosion
Flash floods
Created at base of mountains due to accelerated run-off and reduced absorptive capacities of soil.
Overgrazing
Reduces groundcover
Increases run-off.
Sediment from erosion
Settles to the river bottom.
Gradually raises water level.
Increased urbanization
Prevents ground absorption of water which increases run-off and contributes to flash flooding.
Flood plain
Flood plain
Attracts poor urban dwellers because of inexpensive land values.
Flood plains in rural areas
Attract farmers because of fertile land.
Delta

unreliable for flood protection purposes due to the varying source of the contributing river tributaries.

Coastal floods — Some flooding is associated with tropical cyclones (also called hurricanes and typhoons). Catastrophic flooding from rainwater is often aggravated by wind-induced storm surges along the coast. Salt water may flood the land by one or a combination of effects from high tides, storm surges or tsunamis. (See the chapters on tsunamis and tropical cyclones for more information.) As in river floods, intense rain falling over a large geographic area will produce extreme flooding in coastal river basins.

Q. Is your community or country susceptible to flooding? What types?

A. ______________________________

____

____

____

____

____

____

How do humans contribute to flooding?

Floods are naturally occurring hazards. They become disasters when human settlements occupy the floodplain. Population pressure is now so great that the risks associated with floods have been accommodated because of the greater need for a place to live. In the United States, for example, billions of dollars have been spent on flood protection programs since 1936. In spite of this the annual flood hazard has become greater because people have moved to and constructed upon flood plains faster than the engineers can design better flood protection.

Floods are naturally occurring hazards. They become disasters when human settlements occupy the floodplain.

Increase in population combined with poor resource management have resulted in new types of flooding. Conversion of forests in the catchment area to pasture and arable land means that less water is held in the upper reaches of the catchment basin, and the increased runoff water flows rapidly to the plains, with the effect of more frequent, unexpected and severe flooding.

Another type of flood becoming more common is *urban* flash flooding. Buildings and roads cover the land preventing infiltration so that rainwater runs over the impervious surfaces forming artificial streams. Inattention to maintenance of drainage systems, especially after long dry spells when dust, debris and overgrown vegetation have blocked natural water flow, can accentuate the degree of flash flooding.

General characteristics

Floods may be measured and analyzed by the following criteria:

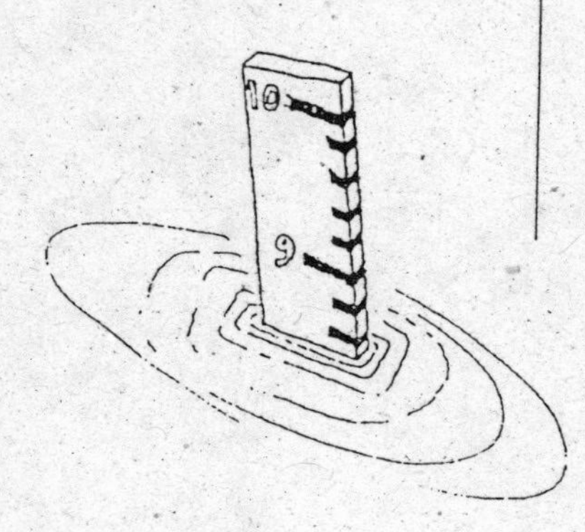

Depth of water — Building foundations and vegetation will have different degrees of tolerance to being inundated with water.

Duration — Damage or degree of damage to structures, infrastructure and vegetation is often related to the length of time of inundation by water.

Velocity — Dangerously high velocities of flow may create erosive forces and hydrodynamic pressure which may destroy or weaken foundation supports. These may occur on the floodplains or in the main river channel.

Rate of rise — Estimation of the rate of rise and discharge of a river is important as a basis for flood warnings, evacuation plans, and zoning regulations.

Frequency of occurrence — The cumulative effects and frequency of occurrence measured over a long period of time will determine what types of construction or agricultural activities should take place on the floodplain.

Seasonality — Flooding during a growing season may completely destroy crops while cold weather floods from snow melts may seriously affect the functioning of a community.

Predictability

Flooding, resulting from precipitation or snow melt in the catchment system, or from upstream flooding, is predictable from about 12 hours to as much as several weeks ahead.

Riverine flood forecasting estimates river level stage, discharge, time of occurrence, and duration of flooding, especially of peak discharge at specific points along river systems. Flooding, resulting from precipitation or snow melt in the catchment system, or from upstream flooding, is predictable from about 12 hours to as much as several weeks ahead. Forecasts issued to the public result from regular monitoring of the river heights and rainfall observations. Flash flood warnings, however, are dependent solely on meteorological forecasts and a knowledge of local geographical conditions. The very short lead time for the development of flash floods does not permit useful monitoring of actual river levels for warning purposes.

For comparison with previous flood events, and conversion to warning information, assessment of the following elements should be included: flood frequency analysis, topographic mapping and height contouring around river systems with estimates of water holding capacity of the catchment area, precipitation and snow melt records, soil filtration capacity, and (if in a coastal area) tidal records, storm frequency, topography, coastal geography and breakwater characteristics.

An effective means of monitoring floodplains is through remote sensing techniques such as Landsat. The images produced by the satellites can be interpreted and used to map flooded and flood-prone areas. Other efforts to improve forecasting are being implemented by UN organizations such as the World Meteorological Organization using World Weather Watch and Global Data Processing Systems. These systems are strategic when flood conditions exist across international boundaries. The great majority of river and flash floods forecasts, however, depend on observations made by national weather services for activation of flood alert warnings.

Vulnerability

At notable risk in flood plain settlements are buildings made of earth or with soluble mortar, buildings with shallow foundations or non-resistant to water force and inundation. Infrastructural elements at particular risk include utilities such as sewer systems, power and water supplies, machinery and electronics belonging to industry and communications. Of great concern are food stocks and standing crops, confined livestock, irreplaceable cultural artifacts, and fishing boats and other maritime industries.

Other factors affecting vulnerability are lack of adequate refuge sites above flood levels and accessible routes for reaching those sites. Similarly, lack of public information about escape routes and other appropriate response activities renders communities more vulnerable.

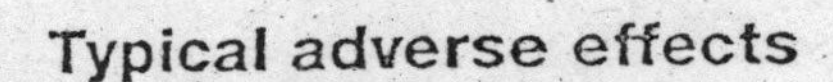

Typical adverse effects

Physical damage

Structures are damaged by a) force of impact of flood waters on structures b) floating away on rising waters c) becoming inundated d) collapsing due to undercutting by scouring or erosion and e) damage by water-borne debris.

Damage is likely to be much greater in valleys than in open, low-lying areas. Flash floods often sweep away everything in their paths. In coastal areas, storm surges are destructive both on their inward travel and again on the outward return to the sea. Mud, oil and other pollutants carried by the water are deposited and ruin crops and building contents. Saturation of soils may cause landslides or ground failure.

Casualties and public health

Currents of moving or turbulent water can knock down and drown people and animals in relatively shallow depths. Major floods may result in large numbers of deaths from drowning, particularly among the young and weak but generally inflict few serious but non-fatal injuries requiring hospital treatment. Slow flooding causes relatively few direct deaths or injuries, but often increases occurrences of snake bites.

Major floods may result in large numbers of deaths from drowning, particularly among the young and weak but generally inflict few serious but non-fatal injuries requiring hospital treatment.

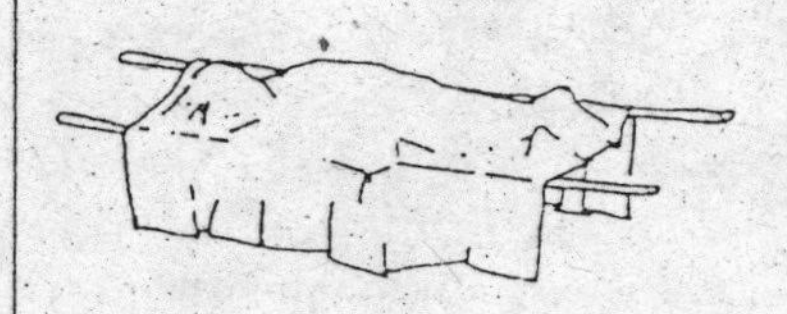

Endemic disease will continue in flooded areas, but there is little evidence of floods directly causing any large scale additional health problems apart from diarrhea, malaria and other viral outbreaks eight to ten weeks following the flood.

Water supplies

Open wells and other groundwater supplies may be contaminated temporarily by debris carried by flood waters or salt water brought in by storm surges. They will, however, only be contaminated by pathogenic organisms if bodies of people or animals are caught in the sources or if sewage is swept in. Normal sources of water may not be available for several days.

Crops and food supplies

An entire harvest may be lost together with animal fodder resulting in long-term food shortages. Food stocks may be lost by submersion of crop storage facilities resulting in immediate food shortages. Grains will quickly spoil if saturated with water even for a short time.

Most agricultural losses result from the inundation of crops. Susceptibility to inundation depends on the type of crop and duration of flooding. Some crops, such as taro are quickly killed by relatively small amounts of flood water. Others may be able to resist submersion but may die eventually if large amounts of standing water stagnate as in the 1988 Bangladesh flood.

The majority of deaths and much of the destruction created by floods can be prevented by mitigation and preparedness measures.

Large numbers of animals, including draught animals, may be lost if they are not moved to safety. This may reduce the availability of milk and other animal products and services, such as preparation of the land for planting. These losses, in addition to possible loss of farm implements and seed stocks, may hinder future planting efforts.

Floods bring mixed results in terms of their effects on the soil. In some cases, land may be rendered infertile for several years following a flood due to erosion of the topsoil or by salt permeation in the case of coastal floods. Heavy silting may either have adverse effects or may significantly increase the fertility of the soil.

In coastal areas where fish provide a source of protein, boats and fishing equipment may be lost or damaged.

On the positive side, floods may flush out pollutants in the waterways. Other positive effects include preserving of wetlands, recharging groundwater, and maintaining the river ecosystems by providing breeding, nesting, and feeding areas for fish, birds and wildlife.

Application of remote sensing data to flood prone areas:

Honduras coastal plain — In September 1974, the coast of Honduras was flooded by Hurricane Fifi. The government of Honduras requested assistance from OAS/DRDE for mapping of the floodplain for use in an integrated development study to rebuild the coastal areas. Using Landsat data sheets, pre-flood and post-flood imagery was overlapped to show the areas susceptible to flooding and where prevention and mitigation measures and land use zoning would be necessary.

Pilcomayo River Floodplain: Recurrent flooding along the Pilcomayo River in south-western Paraguay, prompted the Paraguayan Government to draw up a hazard map of the floodplain. Several different maps such as those showing desertification and soils were combined with Landsat maps to draft floodplain boundaries and hazard zones, using overlay mapping. The maps showed the changes in the river channel over time and served to demonstrate the highly dynamic nature of the floodplain, indicating continuous need for floodplain monitoring. Further analysis was used to assess flood hazard vulnerability.

Possible risk reduction measures

The majority of deaths and much of the destruction created by floods can be prevented by mitigation and preparedness measures. The first step involves identifying vulnerable elements by preparation of a flood hazard map and then integrating that information into a plan for preparedness and development. A strategy might combine regulation of land in the floodplains with flood control measures. Planners may seek contribution from a variety of disciplines to assess risk, the level of acceptable risk, and viability of proposed activities. Information and assistance may be obtained from different sources ranging from international agencies to the community level.

Mapping of the floodplain — Floods are normally described in terms of statistical frequency using the 100 year flood plain event parameters for flood mitigation programs. The 100 year flood plain describes an area subject to a 1% probability of a certain size flood in a given year. Depending on the degree of acceptable risk that is selected for an evaluation, other frequencies may be chosen such as 5, 20, 50, or 500 year floodplains.

The basic map is combined with other maps and data to form a complete picture of the floodplain. Other inputs include frequency analysis, inundation maps, flood frequency and damage reports, slope maps and other related maps such as land use, vegetation, population density and infrastructural maps. In some developing countries, obtaining extensive long term information may be difficult. Remote sensing techniques provide an alternative to traditional techniques of floodplain mapping and can be equally or more cost effective as they allow estimates of data otherwise requiring labor intensive collection methods, as in hydrology studies over extensive areas.

Multiple hazard mapping — Floods often cause, occur in conjunction with, or result from other hazards. A multiple hazard map, known as a composite, synthesized or overlay map, serves to highlight areas vulnerable to more than one hazard. It is an excellent tool for designing a multiple hazard mitigation and emergency plan. It may, however, not be adequate for site-specific, hazard specific engineered activities.

Land use regulations ensure that flood risks are not made worse by ill-conceived new land uses.

Land use control — The purpose of land use regulations is to reduce danger to life, property and development when high waters inundate the floodplains or the coastal areas. Land use regulations ensure that flood risks are not made worse by ill-conceived new land uses. Of particular concern are regions of urban expansion. The following elements should be addressed.

1. **Reduction of densities:** In flood prone areas, the number of casualties is directly related to the population densities of the neighborhood at risk. If an area is still in the planning stages, regulation of densities may be built into the plan. For areas already settled, especially squatter settlements, regulation of density can be a sensitive issue and would have to address the socioeconomic implications of resettlement. Unfortunately, many situations exist where dense unplanned settlements are located on floodplains. Planners must incorporate measures to improve sites and reduce vulnerability.
2. **Prohibiting specific functions:** No major development should be permitted in areas found to be subject to flooding once every 10 years on average. Areas of high risk can be used for functions with a lower risk potential such as nature reserves, sports facilities and parks. Functions with high damage potential such as a hospital are permitted in safe areas only.

Figure 2.2.2
Schematic flood plain regulation map

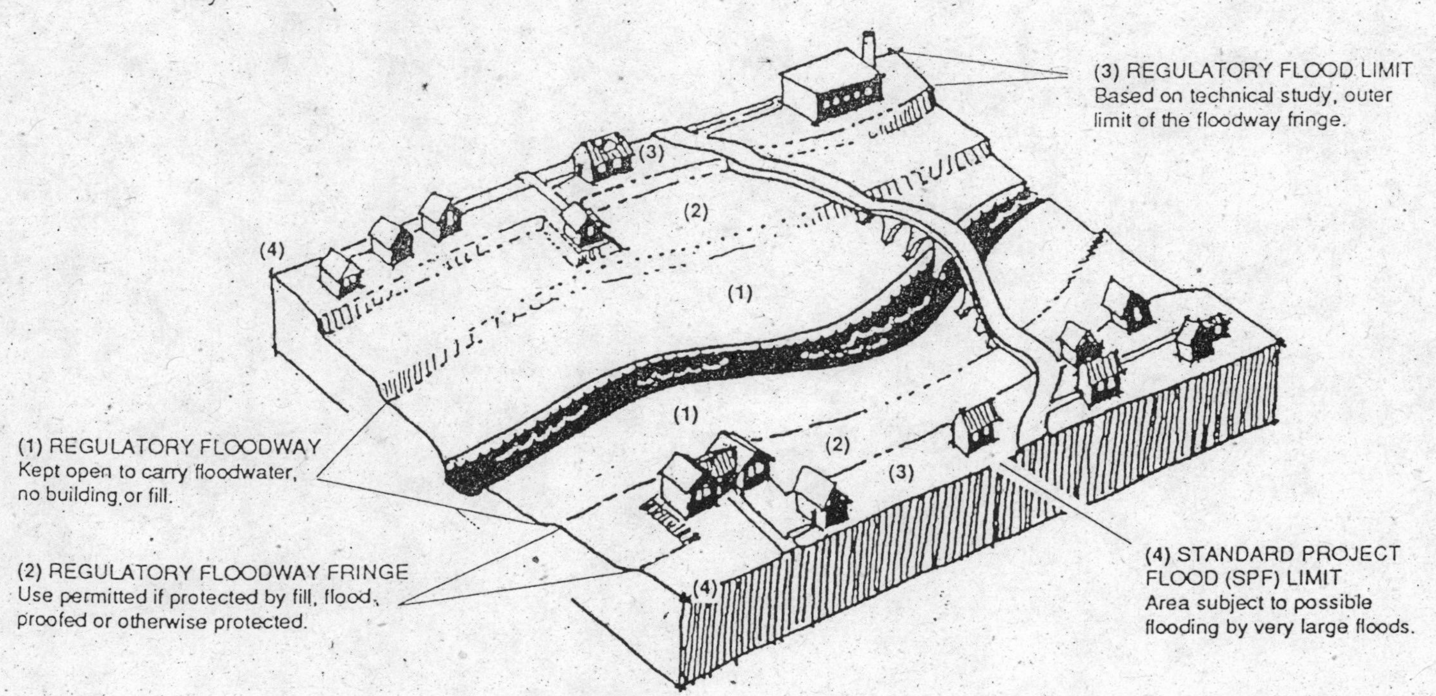
(3) REGULATORY FLOOD LIMIT
Based on technical study, outer limit of the floodway fringe.
(3)
(2)
(4)
(1)
(1)
(2)
(3)
(4)
(1) REGULATORY FLOODWAY
Kept open to carry floodwater, no building or fill.
(2) REGULATORY FLOODWAY FRINGE
Use permitted if protected by fill, flood proofed or otherwise protected.
(4) STANDARD PROJECT FLOOD (SPF) LIMIT
Area subject to possible flooding by very large floods.

3. **Relocation of elements that block the floodway:** In addition to the obvious danger of being washed away, buildings blocking the floodway may cause damage by trapping floodwaters which then overflow into formerly flood free zones.
4. **Regulation of building materials:** In certain zones wooden buildings and other light structures should be avoided. In some cases, mud houses are permitted only if flood protection measures have been taken.
5. **Provision of escape route:** Neighborhoods should have clear escape routes and provision of refuge areas on higher ground.

For developing countries with intensively utilized floodplains, considerable political influence in conjunction with the cooperation of the communities may have to be employed.

Other preventative strategies include:

- the acquisition of floodplain land by developmental agencies, perhaps by swaps that provide alternatives for building sites
- establishment of incentives (loans or subsidies, tax breaks) to encourage future development on safer sites using safer methods of construction
- diversification of agricultural production such as planting flood resistant crops or adjusting the planting season; establishing cash and food reserves
- reforestation, range management and animal grazing controls to increase absorption (see chapters on deforestation, desertification)
- construction of raised areas or buildings for use as refuge if evacuation is not possible.

Figure 2.2.3

Schematic flood plain regulation map

Natural Hazards, Disaster Management Center, 1989.

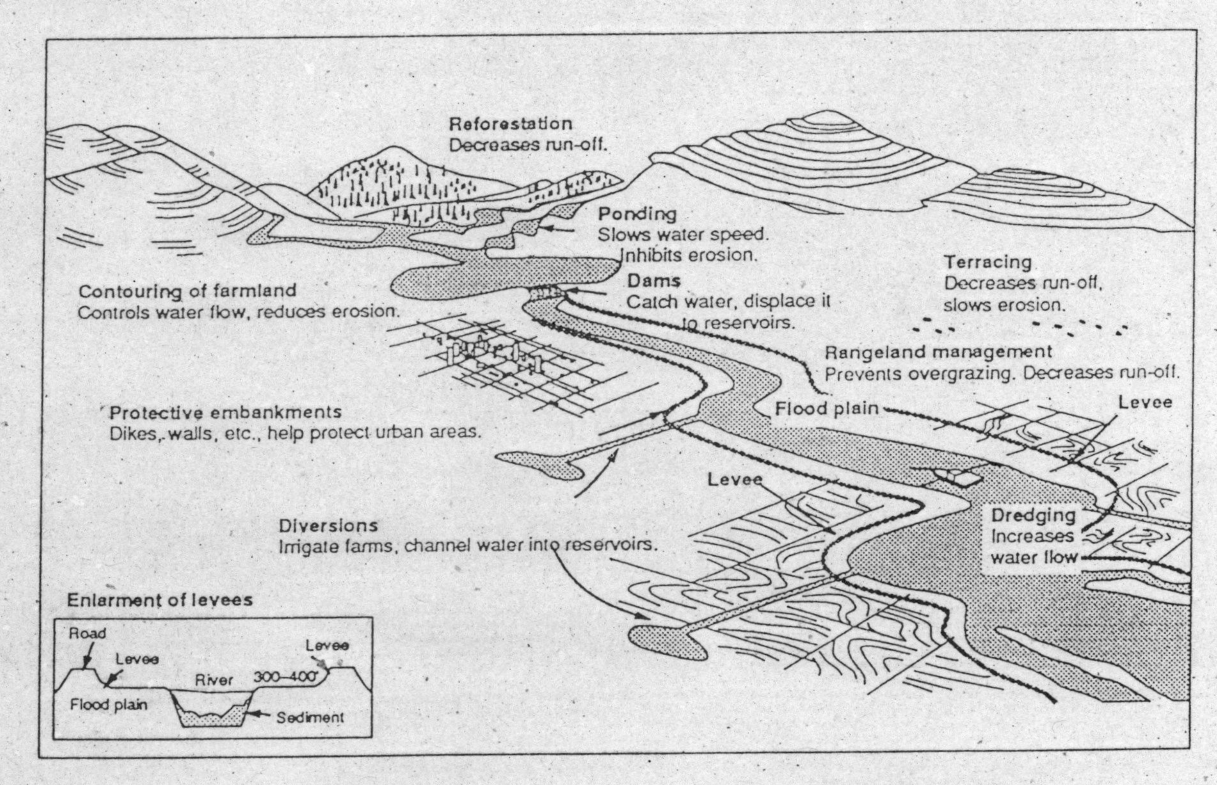

Flood control

As mentioned above, land use controls will be of limited use in already developed floodplains. Yet, changes must be implemented to reduce a community's vulnerability to flood damage. For developing countries with intensively utilized floodplains, considerable political influence in conjunction with the cooperation of the communities may have to be employed. The most commonly used options are:

Existing channel improvements — Deepening and widening the river bed are methods to increase capacity and thus reduce the area of the floodplain.

Diversion and relief channel construction — New channel construction may be a feasible alternative relative to the cost of moving a settlement. Several options exist for channel construction such as open grass-lined channels, concrete or rock lined channels. Great care must be exercised in the design and construction of diversion channels because of the possible environmental impacts and necessary safety features. Costs may be prohibitive for such highly engineered works.

Dikes and dams — These facilities are capable of storing flood water and releasing them at manageable rates. Again, careful engineering is required to anticipate maximum flood levels. If exceeded, the damage caused may be much higher than if the facility had not been built. Dams and other retention facilities may give the public a false sense of security if not properly engineered and constructed.

Flood-proofing — Individual property owners may reduce the risk of damage by strengthening buildings to a) resist the water's force and b) retain integrity when inundated with water. Newly constructed buildings should have foundations which are not susceptible to scouring .

Protection against erosion — An important element of flood defense is protection against erosion. Streambeds should be stabilized with stone masonry or vegetation especially near bridges.

Site improvement — The elevation of sites can be an effective option for individual or group dwellings.

Q. *What are some possible risk reduction measures which may be used in regard to flood hazards?*

A.

1. 'Technical information on floods' taken from: *Introduction to Hazards*, UNDP Disaster Management Training Programme.

Why should emergency preparedness and response be integrated with development?

Purpose

This activity allows participants to develop their understanding of how different pressures obstruct attempts to integrate relief and development action. It also builds participants' abilities to argue for integration.

Practical considerations, long-established practices and organisational dynamics often stand in the way of integrating everyday preparedness and response with development.

In a process of developing arguments, players have to focus clearly on the implications of integration or non-integration. They need to think of clear strategies, and practise their skills in presenting convincing arguments. This also involves exploring ATTITUDES that hinder or promote a process of integration.

Procedure

The activity takes the form of a role play in which different role players representing various interests argue for or against the integration of specific development actions into drought relief.

Time

- 2½ hours

Materials

- briefing sheets for different roles (see resources)
- pens / paper for observers

Process

Introduction

1. Explain the purpose of the activity and outline the process.

2. Ask participants to have a brief buzz, with the person sitting next to them, about reasons for /against an integration of relief and development activities.

3. Initiate and facilitate a brainstorm. Remind participants of the rules and purpose of a brainstorm. Record all key words.

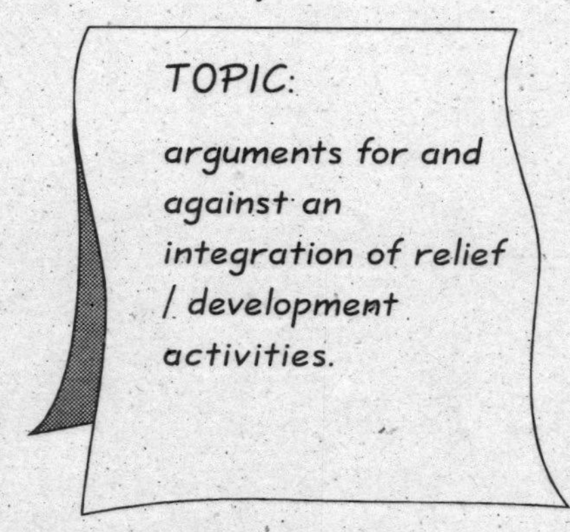
TOPIC:
arguments for and against an integration of relief / development activities.

4. Display the keywords and suggest that participants might want to refer to them in the following activity.

Participant Action

1. Introduce the role play
   Ask for volunteers to take on the roles of finance officer, health programme officer, disaster reduction officer, head of programmes and agency director.

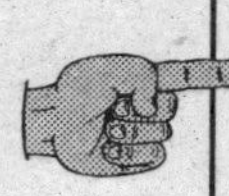

Depending on the number of participants you may want to allocate two (or more) players to each of the roles and run two (or more) simultaneous role plays. This allows for maximum participation.

2. Ask remaining participants to take on the role of observers.

3. Briefly outline the process:
   (i) Groups will work for 45 minutes.
   (ii) Observers will have 10 minutes to develop their reports.
   (iii) Reports will be presented in plenary.
   (iv) This will be followed by a plenary discussion around key issues as they arise from reports and discussion about the process.
   (v) The process will conclude with a discussion around a number of questions.

4. Distribute briefing sheets, read through them and ask participants to prepare arguments. Allow 5-10 minutes.

5. Begin the role play. Point out that you will move from group to group and be available to answer questions of clarification should the need arise.

6. At the end of 45 minutes ask observers to prepare their reports. Ask role players to move back into plenary.

Review and Discussion

1. Review the role play.
   Ask participants questions such as the following:

   ? How did they feel in their roles?

   ? What did they like / dislike about playing the role they were assigned?

   ? What did they find difficult /easy about their roles?

   ? What attitudes were displayed? Are these realistic?

   Indicate that this is the end of the role play and they will now assume their own identities. Ask players from each group to address each other by name and shake hands.

2. Ask observers to join the group; point out that in the following you will examine arguments for and against integration and that these arguments should be viewed as independent from the role players. Encourage observers to refer to roles rather than players in their reports. (Talk about 'the health officer' etc, and not the person who took on that role.)

3. Take reports from observers and record main key arguments on flipchart.

4. Facilitate a discussion around those arguments:
   - ? Which of these are based on 'myths'? Which are real?
   - ? How can arguments against integration be countered?
   - ? What were solid arguments for integration? What made them so convincing? Why were they effective?
   - ? What other arguments could be put forward (these might include some suggested during the brainstorm but not developed further in the role plays)?

Some good arguments for integrating development and relief activities

1. In repeatedly drought-stricken areas, a community-based health programme will have greater credibility with the targeted communities if it addresses recurrent problems caused by drought conditions (e.g. how relevant is it to offer HIV/AIDS education to community members who have no food or water today?)

2. Health programmes that integrate drought considerations during their planning phase are more likely to be "drought proofed' when drought does strike — thus more sustainable, and able to continue offering a service when the community needs it most.

3. The visibility that accompanies a relief programme can bring welcome publicity to ongoing but low-profile development activities.

4. When relief activities are implemented vertically, without careful integration with ongoing village, community or district infrastructures, they are very fragile and non-sustainable after the relief programme ends. If a new relief programme is required 2-3 years later, it is more difficult to resuscitate a mechanism that has been in "hibernation" than one that rests on ongoing community programmes and structures.

5. The time to launch long-term recovery and mitigation activities is right at the start of a relief operation. Although these types of activities are almost all developmentally-orientated, communities are much more receptive towards them when they are dealing with the full impact of a drought — rather than later.

5. Suggest the importance of transferring these insights into participants' work in the field/their organisations: how could they apply what they learnt during the role plays/discussion in their work? What are the obstacles and how could those be overcome?

6. Ask a participant to summarise the main points and key issues. Suggest participants may want to take those forward in their report-backs to their organisations.

Notes

Resources

ROLE PLAY

Arguments for integrating EPR with development

Scenario (for all participants)

You are participating in a meeting called by the Disaster Reduction Officer in order to discuss a proposed programme of drought action in the region. You are all members of an active NGO - working at Headquarters level. The agency director will chair the meeting. Also present are the health officer, disaster reduction officer, head of programmes & finance officer.

This year, once again, a drought has been declared that affects several districts in your country. Your organisation has been assigned two districts with 80,000 people to assist. The officer in charge of disaster reduction activities in your agency has called for a meeting on how to proceed. S/he has supervised drought relief efforts twice before, and now is deeply committed to using this opportunity to reduce the risks associated with repeated drought in the two districts. As your agency has a limited community-based health care programme in this area, s/he would like to link health and drought related risk reduction activities more closely.

The purpose of the meeting is to reach an agreement on whether this is feasible, and how the operation should be structured. In each district there is a project officer for health, and a number of health volunteers, who have had basic training in primary health care. UNICEF has offered some funding for this effort (that is the drought programme) and so has the World Food Programme. However, in the disaster reduction officer's opinion this support is not enough, and she would like to access some of the health programme's resources for the 12 month risk reduction programme. However, the health officer has already designed her own plan for this year for these districts and is unwilling to divert from the established plan. She is also concerned that the donors might react negatively.

You have 45 minutes to decide on how this issue will be resolved.

Individual briefing sheets for role players

Finance Officer

You have recently joined the NGO, having worked for a borehole drilling company for eight years. Because of your background and familiarity with water resource rehabilitation, you are open to broadening the drought operation to include water and health components.

At the same time, you are a bit worried about how you would account for monies diverted from the health programme to co-fund the drought operation and how this would be safely monitored - especially if the monies were disbursed at district level. You have a great sense of humour.

Health Programme Officer

You are absolutely devoted to CBHC & PHC and have worked hard to generate funds for your programme. You have worked for this NGO for five years - and have virtually built the health programme from zero. Now - there are trained health field officers in each district - all with motor bikes, and you feel you have wonderful infrastructure to mobilise. You have designed a careful plan for the next year, and are determined to prove to the director and your donors that you can make it work. Your priorities are HIV/AIDS education for women and sanitation. You have always thought that your organisation's work in drought situations was limited to food relief, and are unwilling to divert your health workers, transport and donor monies away from your designed plan to reduce drought-related risk.

Disaster Reduction Officer

You have worked for this NGO for a number of years. You have wanted to build your organisation's capacity in disaster reduction projects for several years now, but are always getting tied up in drought operation & monitoring food relief in the same districts.

This year you have decided that the only way to reduce the repeated need for food relief in the districts assigned, is to have a broader programme with health and water components. In the last drought, the area suffered badly with diarrhoea and dysentery. You know that the money from UNICEF & WFP will not be enough to cover all costs, and see that one strategy would be to involve the health officer and her programme. But you realise she will not divert resources without good reasons. Moreover, the head of programmes and agency director are likely to have difficult seeing the connection between drought and CBHC without a very clear argument, You are prepared for the meeting with diarrhoea statistics from the last drought in the 2 districts (make these up). You are under pressure to get food relief started but don't want to do this before securing a commitment from you colleagues that the operation will be integrated with health.
How will you convince them?

Head of Programmes

You are the institutional memory for this agency, having worked with it for 10 years. Formerly you worked as a field officer and then for three years in charge of disaster relief. You are well liked by all colleagues and are very fair.

However, you have not been able to keep up to date with recent developments in thinking related to disasters and development. You can understand why the government has asked your NGO to do the disaster relief in this drought - but can't quite understand what this has got to do with health. You are worried that involving both health and disaster reduction officer in the same project will present a real headache for supervision and monitoring. However, as you like the disaster reduction officer, you are open to listen to new arguments.

Agency Director

You are the agency director - and have been for 2½ years. You are eager to build your organisation's profile, and view the drought operation as a perfect opportunity. You are preoccupied with sustaining funding for core expenses - don't wish to antagonise donors.

You respect your health and disaster reduction officers very highly. You have a basic understanding of PHC as your brother works in the Ministry of Health. However, your understanding of the spectrum of activities in disasters is quite conventional and limited to more traditional response.

Observer 1

List all good arguments for integrating health and the drought operation to achieve long-term risk reduction.

Observe and list the positive attitudes that support these arguments.

What strategies are used to convince other members of the group?

Observer 2

List all the arguments against integration.

Observe and record the barriers against a change in work policy and the attitude that displays this negative stance.

What strategies are used to maintain the status quo?

How do we integrate risk reduction with emergency response actions?

Purpose

Participants improve their ability to link emergency response action with long-term development planning.

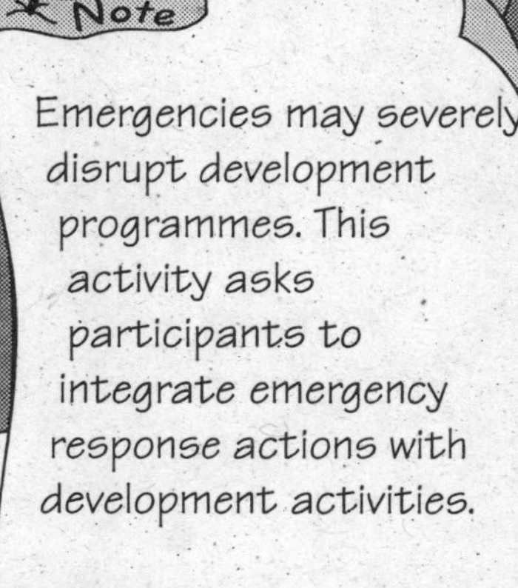

Note
Emergencies may severely disrupt development programmes. This activity asks participants to integrate emergency response actions with development activities.

Procedure

Participants focus on a specific hazard, and list actions on cards. These are matched and arranged to form the basis of plans.

Time

- 1 - 1½ hours

Materials

- blank cards in two colours

Process

Introduction

1. Outline the purpose and procedure of the activity.
2. Select a hazard that is prevalent in Southern Africa, such as drought, AIDS, cholera, flood or civil unrest.
3. Distribute the blank cards; ensure that each participant has at least one of each colour; leave spare cards accessible to participants.
4. Explain that there are two sets of cards: blue for emergency response actions, and yellow for risk reduction measures.

Participant Action

1. Ask participants to think of specific actions in response to an emergency linked to / arising from the hazard. Each participant is to write no more than one action on a blue card; point out that participants may fill in more than one card.
2. Ask participants to think of long-term measures that would reduce the risk of the hazard and write such measures on yellow cards (one measure per card).
3. Collect all blue and yellow cards; ask participants to get into two groups: a blue group and a yellow group. Give each group their pile of cards.
4. Give the following instruction:

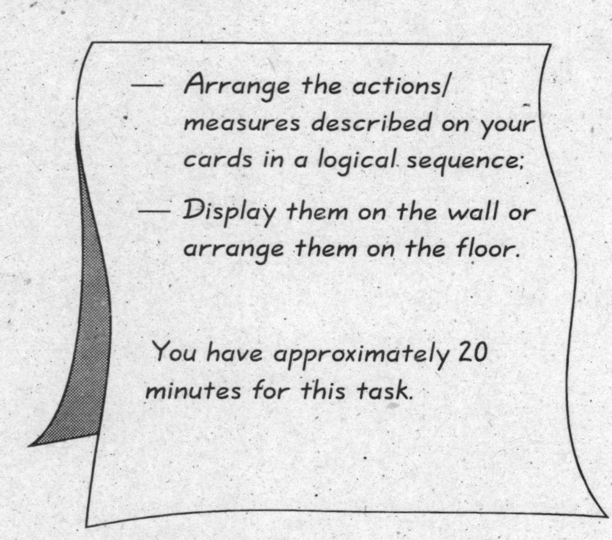

— Arrange the actions/ measures described on your cards in a logical sequence;
— Display them on the wall or arrange them on the floor.
You have approximately 20 minutes for this task.

Review and Discussion

1. In plenary, review the actions and measures listed and the arrangement of cards; ask questions such as the following:
   - ? Are there any important omissions?
   - ? Are any actions or measures inappropriate?
   - ? Have gender planning considerations been taken into account?
   - ? Are the actions and measures in logical sequence?
2. Encourage participants to add or change cards, if necessary.
3. Ask participants to consider who would perform the actions and implement the measures described:
   - ? Which actions are done by an agency, which by the community? Which actions are shared, and who controls the process?
   - ? Who would participate in the process, and in what capacity?
   - ? What skills skills / information are needed to perform the action? Who has those skills?
4. Compare the arrangements and address issues such as the following:
   - ? At which point could emergency and development actions best be integrated?
   - ? Where do they overlap?
   - ? Where are they in conflict?

Participant Action

1. Suggest that in order to affect sustainable development, emergency response actions and risk reduction measures should go hand in hand.

   Ask participants to re-arrange the blue and yellow cards in order to develop one integrated action plan.
2. Clearly identify and mark key steps and identify potential problem areas.

Review and Discussion

1. Sum up by asking participants to suggest how they will apply their learning in the field:
   - ? How could they use the integrated action plan in their work?
   - ? How could they use their understanding of the interrelationship between relief / emergency response and development work?

Notes

How is gender analysis a tool for risk reduction planning?

Purpose

Participants will develop their understanding of how gender analysis is a useful tool for identifying development opportunities in an emergency situation.

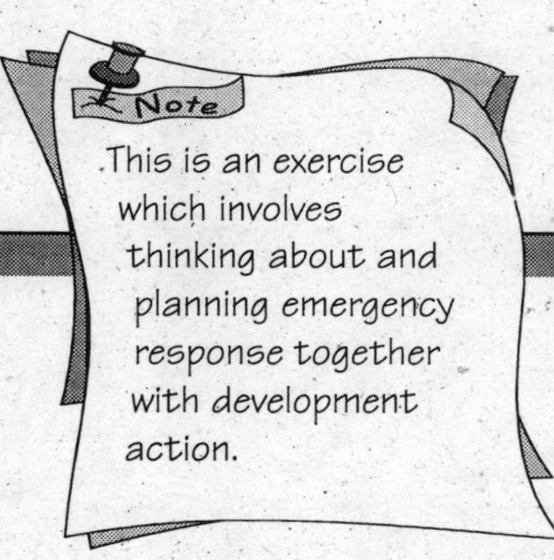

Note
This is an exercise which involves thinking about and planning emergency response together with development action.

Procedure

This is a reading and discussion activity in which participants practise the skills necessary for ensuring a gender-conscious approach to risk assessments.

Time

- 1 - 1½ hours

Materials

- copies of Mary Anderson 'Understanding the disaster development continuum. Gender analysis is the essential tool', for each participant
- discussion questions written on flipchart or typed and copied for each participant

Process

Introduction

1. Outline the purpose and procedure of the activity.

2. Distribute copies of the article and ask participants to spend approximately 15 minutes reading through it.

Participant Action

1. After 15 minutes display and /or distribute discussion questions such as the following:

(i) What arguments does Anderson put forward for a switch from dealing with the symptoms of disasters to addressing the causes?

(ii) Give examples of how people have played a role in creating hazardous situations in communities you know.

(iii) Anderson illustrates how women are vulnerable not because of their sex but because of their gender. Describe the difference!

(iv) Anderson argues that emergency situations can be utilised for addressing both the immediate needs of disaster victims and preparing sustainable development efforts. Respond to the examples she gives and provide examples from your own experience when this goal was realised.

(v) What are the essential tools that enable communities and agencies to recognise and realise such opportunities?

2. Read through the questions aloud, and check for understanding.

3. Ask participants to spend 5 minutes responding to the questions individually. Then ask them to get into small discussion groups.

   Allow approximately 30 minutes for discussions in groups.

Review and Discussion

1. In plenary, encourage brief responses to the issues raised by the discussions.
2. Request responses to the last question, and list the tools on newsprint.
3. Sum up by asking participants to work in pairs, and give the following instruction:
   * working in pairs, take turns in telling each other how you will use a gender analysis for assessing vulnerabilities of both men and women, in your work in the future.

   You will each have 3 minutes to speak.

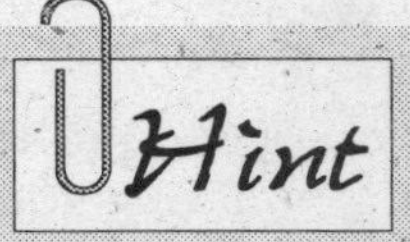

This activity could be followed by asking participants to give examples from their own experiences that illustrate how gender has led to the increased vulnerability of women. What strategies could correct such inequitable delivery of short- and long-term interventions?

Resources

READING

Mary B Anderson

Understanding the disaster-development continuum: Gender analysis is the essential tool[1]

Increasingly, the agencies of the United Nations, the development bureaux of donor nations, and the large number of non-governmental organisations (NGOs) that work in countries striving to achieve development are focusing their attention on understanding the relationships between disasters and development. This focus is motivated by two recent trends. First, the number of disasters worldwide is rising, with an increasing number of people suffering as a result. Recognising this situation, aid workers are expressing growing frustration that they continue to respond only to symptoms rather than addressing the causes of disasters. Second, a reduction in overall aid budgets is apparent in many donor countries today, with an accompanying shift of these shrinking funds away from development programmes and towards disaster response. As a result, both development and relief workers are seeking ways to use available relief funds to meet the emergency needs of disaster victims and, at the same time, support fundamental change towards long-term development.

These two motivations - an urgent need to deal with the causes of disasters rather than only with the symptoms, and the necessity of getting the best possible short-term and long term outcomes from aid funds - are forcing a harder look at the tools that are available for effective planning and programming. One such tool, which can contribute significantly to addressing root causes and which can support effective, efficient and equitable long-term development, is gender analysis.

Causes rather than symptoms

When considering the causes of a disaster, the basic question is: What makes disasters happen? It is now widely acknowledged that disasters occur as a result of human actions and human decisions, rather than as 'acts of God'. A strong wind at sea that does not cause damage to human life or property does not represent a disaster, whereas if that same wind comes on shore where people have built flimsy homes in vulnerable locations it will create a disaster. An earthquake can cause massive death and damage when buildings are weak and preparations inadequate, but another earthquake of the same force can cause little or no damage where building technologies have been developed to withstand tremors and building codes adopted and enforced to ensure that these technologies are used. Increasing flooding occurs downstream from deforested areas or where silting has occurred as a result of erosion. Human agency plays a role in whether or not these natural phenomena - winds, earth movements and rains - do or do not become disasters. The centrality of human action and choices in causing disasters is even more obvious in the growing number of 'complex' emergencies - that is, those disasters that involve both environmental elements and civil conflict.

Because disasters are not brought about solely by natural causes, their impacts are not random. Some individuals and groups become victims while others remain relatively unscathed. The first step in understanding and preparing to deal with root causes is to ar lyze why some people are vulnerable to disasters and others are not.

Vulnerability to disasters can be analyzed in three categories[1]. First, people may be physically vulnerable. They may live in poorly-built houses on land that is susceptible to catastrophe; they may be poor and have few reserves and no insurance to aid recovery if some crisis occurs. Second, people may be socially vulnerable by being marginalised and excluded from decision-making and political processes. Third, people may be psychologically vulnerable if they feel powerless, victimised, and unable to make effective actions for their own security.

How does gender analysis help us understand vulnerability? Gender is certainly not the only determining factor of vulnerability, nor is it always the most important. However, very often an understanding of vulnerability and the development of strategies for overcoming it can be advanced through gender analysis.

It is often said that 'woman are among the most vulnerable'. Why is this so? Women are also strong and capable. They manage and sustain families under the most deplorable conditions. They are producers of a range of goods and services on which the survival of their societies depends. What makes them vulnerable?

In general, around the world, women are poorer than men. Their poverty arises from the roles they are assigned and the limits placed by societies in their access to and control of resources. Women are disproportionately employed in unpaid, underpaid and non-formal sectors of economies. Inheritance laws and traditions, marriage arrangements, banking systems and social patterns that reinforce women's dependence on fathers, husbands and sons all contribute both to their unfavourable access to resources and their lack of power to change things. The health dangers that result from multiple births can contribute to interrupted work and low productivity. Traditional expectations and home-based responsibilities that limit women's mobility also limit their opportunities for political involvement, education, access to information, markets, and a myriad of other resources, the lack of which reinforces the cycle of their vulnerability.

Understanding these linkages through gender analysis makes it clear that women are vulnerable not because it is in their physical nature to be weak but because of the arrangements of societies that result in their poverty, political marginalisation, and dependence on men. As the number of households headed by women increases, worldwide, these causes of vulnerability have broader implications for the dependants in such families.

Furthermore, understanding that vulnerability is a condition caused by human actions and attitudes can provide insights about strategies for addressing vulnerability and thus dealing with the causes, rather than symptoms, of disasters. Poverty-reduction strategies should have as one major focus the reduction of poverty among women and, particularly, among female-headed households. Such efforts must be designed with attention to the educational, locational, time and tradition-based constraints that women encounter.

Gender analysis can also aid the identification of circumstances in which men may be vulnerable. An example of this comes from a refugee camp in Western Ethiopia where many young Sudanese men were gathered who, having walked long distances to escape conscription into armies, were in exceedingly poor health. Food was immediately shipped into this camp in quantities considered adequate to rebuild their health, but morbidity and mortality remained high. Investigation showed that these male refugees were continuing to starve because the food they were given needed to be cooked before it could be eaten, and their gendered roles had precluded their ever learning about food preparation.

Linking Short-term help to long-term outcomes

The second issue faced by aid workers today is the necessity of ensuring that short-term, relief assistance both meets immediate needs of disaster victims and, at the same time, supports their achievement of long-term developmental goals. Too often, relief assistance has increased the dependency of recipient of continuing aid rather than enabled them to move forward toward self-sufficiency. for example, it is now widely recognised that an influx of donated food, deemed necessary for saving people from starvation, can also undermine market prices and, therefore, the incentives of local farmers to plant the next season's crop. Thus, relief aid can contribute to future and spiralling disaster conditions. Less recognised, but also well-documented, is the long-termnegative effect on relief recipients of organisational systems for distributing goods or prioritising needs that are imposed by donors in their anxiety to meet urgent emergency needs efficiently. Approaches that they deny and undermine the existing physical and organisational capacities of recipient groups also undermine and weaken their subsequent abilities to plan, manage and achieve independent self-sufficiency.

Again, gender analysis provides one critical tool for understanding the linkages between short-term aid and long-term outcomes. If it is critical, when providing emergency assistance, to work with rather than for disaster victims in order to ensure positive long-term impacts, then it is equally critical to identify the capacities of the recipients concerned, because it is these capacities that must be supported if long-term development is to be achieved, The gender roles ascribed to men and women mean that they have different physical, social and psychological capacities in any given context. the scarcity of aid resources makes it even more important to target and tailor assistance to fit local realities.

To return to the situation described above, when aid workers identified the cause of continuing hunger among the young Sudanese men in ethiopian refugee camp, they were able to organise the ten per cent of the population who were women with cooking skills, to teach the men how to cook. Recognition of and support for existing capacities associated with gendered roles may make the difference between a programme's effectiveness or failure. For example, efforts to arrange water supplies for disaster victims and development project participants alike have, often, succeeded or foundered depending on whether they took account of women's involvement in water collection and usage, and whether they are supported, or failed to support, women's capacities to manage and maintain water pumps and other equipment. Similarly, experience has provided too many examples of programmes that have issued drought-resistant seeds and provided technical assistance about new farming methods to male members of disaster-affected groups, to enable families to replant and increase food production, these technologies have not been adopted.

In all societies, men and women experience different vulnerabilities and have different capacities as a result of their gendered roles. Sometimes these roles are very different and rigid; sometimes they are overlapping and fluid. In either case, the failure to identify gendered roles and to plan programmes with them consciously in mind has resulted in the inequitable delivery of disaster relief assistance, and inadequate attention to the potential long-term outcomes of short-term interventions. The tool of gender analysis is a powerful one for accurately diagnosing opportunities and constraints in any programme situation, and for identifying more effective strategies for delivering emergencyassistance so that it supports long-term development for women and men, and girls and boys.

1. Anderson M.B. Understanding the disaster-development continuum: Gender analysis is the essential tool, in *Focus on Gender*, Vol. 2, No.1, February 1994.

Why must gender considerations be essential to emergency preparedness and response planning?

Purpose

This activity demonstrates the need to integrate gender considerations into risk assessment and emergency preparedness and response (EPR) planning.

Note

In this simulation participants practise how to identify elements most at risk in a flood situation. They explore and discuss the critical link between vulnerability assessment and EPR planning.

Procedure

The initial simulation asks participants to shift perspectives and assume the role of vulnerable members of a community during a flood emergency. As vulnerable households they experience the emergency, and then reflect critically on how decisions about actions impact on the lives and livelihoods of different community members.

Time

- 1 hour

Materials

- large plastic sheet draped over objects which have been set up in advance to create a 'landscape' of hills surrounding a river or lake
- different kinds of seed pods or bits of wood, clearly marked to serve as 'houses'
- a container of water
- individual 'household' cards (see resources)
- three sets of different coloured 'tokens' or cards
- a key-chart which outlines the meaning of the tokens written on flipchart

Process

Introduction

1. Outline the purpose and procedure of the activity.

2. Point out that participants will work as households, either alone or in pairs;
   — assign roles and give each household a 'household' briefing card.

3. Give each household a seed pod or bit of wood as a 'house'.

4. Ask one participant to act as 'risk assessment officer'; this involves looking after the tokens and when the time comes; exchanging them for briefing cards.

Participant Action

1. Ask participants to move to the model and place their 'houses' in the valley around the river/lake, or follow the instruction given on their card. Ask them to remember clearly where they situated their houses.

2. Pour water on the model; ensure that the majority of houses in the 'floodplain' tumble / get submerged.

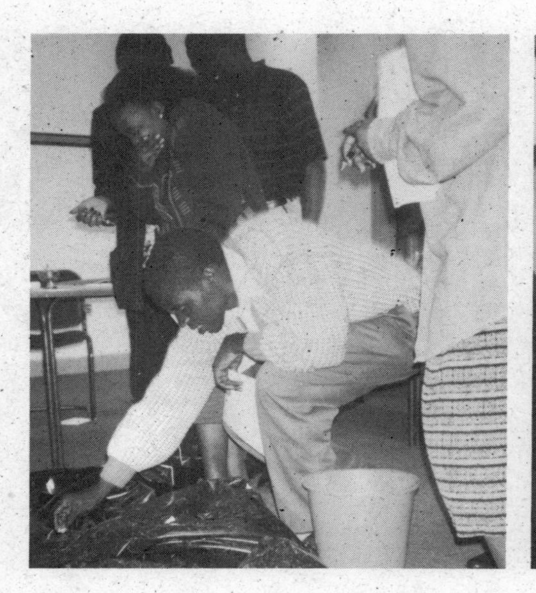

3. Announce that the flood has happened; participants are now asked to conduct an assessment and determine which households are most at risk and were most urgently affected by the flood.

   Explain that there are three categories of affected households: green, blue and pink; point to the key and explain what each category means.

KEY FOR TOKENS

green tokens = house has survived
people have survived

blue tokens = house has gone
people have survived

pink tokens = house is gone
people have died

4. Outline the following procedure for the assessment: taking turns, each household will read out the description on the household briefing card. All participants will then decide into which category ('blue', 'green' or 'pink') the household falls.

5. Once the decision has been taken, ask the volunteer to exchange the household card for a green, blue or pink token, and to affix the briefing card to the key-chart.

6. Proceed to classify each household in this manner.

7. Ask participants to assess the information gathered:

   ? How many households have no tokens? (1)

   ? How many households have green tokens? (2)

   ? How many households have blue tokens? (4)

   ? How many households have pink tokens? (4)

8. Assess the members of households in terms of men/women: ask

   ? How many members of the pink token holders are women? (3)
   3 out of 4 — what percentage is that? (75%)

   ? How many members of the blue token holders are women? (1)
   1 out of 4 — what percentage is that? (25%)

   ? How many members of the green token holders are women? (1
   1 out of 2 — what percentage is that? (50%)

   Request participants to analyse this information: what does it mean in terms of who is most at risk in emergencies such as this one?

Review and Discussion

1. Encourage discussion around the following questions:

   ? Why have more women than men lost their homes and lives?

   ? Did they choose to live near the river bed?

   ? Why did some of the women remain in their homes, and consequently lose their lives?

   ? How did men respond to the emergency situation?

2. Point out how women have to make choices that pull them in different directions, as they consider family members, means of livelihood and community structures/ issues.

   The different pressures put on women correspond to their various roles as reproductive, productive and community workers.

3. Write the terms 'reproductive role', 'productive role' and 'community managing role' on newsprint and ask participants to briefly define what each one means; encourage them to give examples of each as illustrations.

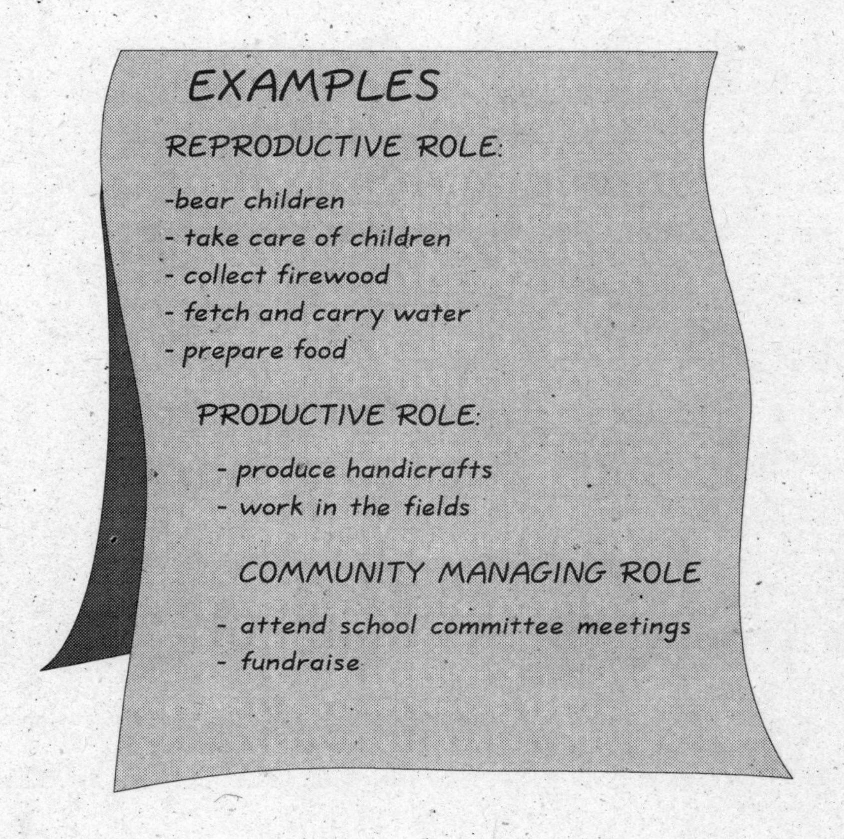

EXAMPLES
REPRODUCTIVE ROLE:
-bear children
- take care of children
- collect firewood
- fetch and carry water
- prepare food
PRODUCTIVE ROLE:
- produce handicrafts
- work in the fields
COMMUNITY MANAGING ROLE
- attend school committee meetings
- fundraise

4. Ask participants to read out examples from their household cards that illustrate the forces attacking women's reproductive, productive and community roles in this simulation.

5. Take the example of the woman who did not survive the flood because she believed that survival of a father means the survival of the family, and her duty was to stay with the father even if the husband left.

   — Initiate a discussion on how cultural and religious traditions sometimes increase women's vulnerability and put them at risk in emergencies.

   — Ask participants for other examples from their culture that illustrate this point; for example: are there specific behaviour-codes or dress-codes, or rules regarding food for women that impose a risk in times of emergencies?

6. Pose the following questions: record the responses on newsprint.

What are some recommendations for EPR plans with regard to reducing the vulnerability of women?

How would identifying people most at risk improve the effectiveness of an EPR plan?

7. Sum up the activity by asking participants to review the lessons learnt.

Learners' Responses

Some of the following quotes from the SADMTP highlight the issues around gender, risk and emergencies thrown up by this simulation. You may want to introduce these quotes to further stimulate discussion.

"This simulation illustrates the old saying: 'Each man for himself and God for us all'."
"Each one to himself - instead of a community effort."
man to man: "Why did you leave by yourself?"
other man: "Anybody would think of themselves first."
women: (rising in protest)
Question: "Is that the same for women?"
man: It's the same for everybody."
woman (protesting): "You're not a woman. Let us answer this question for ourselves."

Notes

ROLE PLAY

Household briefing cards

(These were developed for the participants on the SADMTP Course. Depending on the number of participants on the course you may wish to develop further household descriptions.)

Person 1:
You are a 35 year old widowed woman who is caring for your elderly father. You are so poor that you live right on the river bank.

Person2:
You are a seventy year old man who is disabled and dependent on his daughter (person 1).

Person 3:
You are a merchant in the community, You own a store. You live in a 2-storey concrete house on a hill, some distance from the river.

Person 4:
You are the merchant's wife.

Person 5:
You are a 30 year old single man who owns a bicycle repair business. You take a few important tools out of your tool box and swim across the river.

Person 6:
You are a 20 year old woman. You are married to person 7.

Person 7:
You are a 20 year old construction worker, married to person 6.

Person 8:
You are a 45 year old widow. Last week you spent your life savings on a sewing machine. You won't leave it behind.

Person 9:
You are a 20 year old man who lives with his sister. You are visiting friends when the flood comes. You have to make the choice between fleeing across the river or going back to see if your sister is safe. What do you do?

Person 10:
You share a house with person 9. You are a 20 year old single mother who is eight months pregnant and who has a 2 year old toddler. Can you get across the fast flowing river?

How do we incorporate gender into risk reduction planning?[1]

Purpose

Participants will develop their ability to apply gender analysis tools to assess the impact of mitigation projects on communities.

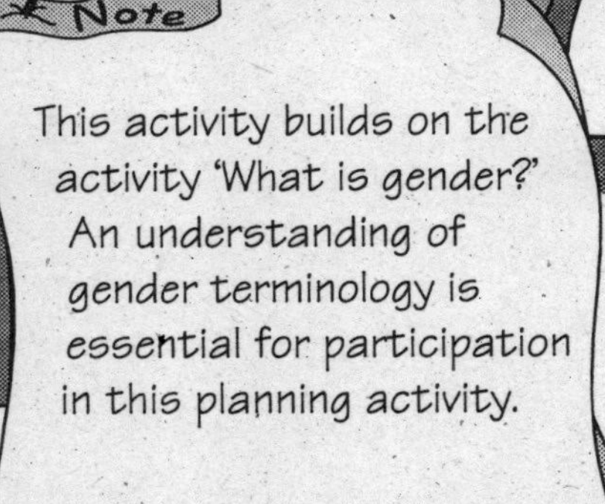

Note
This activity builds on the activity 'What is gender?' An understanding of gender terminology is essential for participation in this planning activity.

Procedure

Participants work in groups and analyse given case scenarios, using a set of questions as a guideline.

Time

- 1½ hours

Materials

- copies of case scenarios and instructions for each participant (see resources)

Process

Introduction

1. Outline the purpose and procedure.

2. Introduce the activity by pointing out that participants will need a common language in order to complete the following activity. Re-establish an understanding of terminology by asking participants to define each of the following terms:
   - gender planning;
   - productive, reproductive, community managing/politics roles;
   - personal and strategic gender needs.

3. Ask participants to form discussion groups. They will work with either case study A or case study B.

Participant Action

1. Hand each participant a copy of either case scenario (A), or case scenario (B), and an instruction sheet.

2. Read through the instruction sheets and check for understanding; clarify where necessary.
   Remind participants of the time limit (40 minutes)

3. Monitor the group work and assist where necessary.

Review and Discussion

1. Manage the report-back process and questions of clarification.

2. Initiate discussion around the following questions:
   - **?** What are the similarities and differences between the case scenarios?
   - **?** How did the agencies' failure to consider gender impact on the women and men in their various roles?
   - **?** Which gender needs did the interventions fail to meet?

3. Remind participants that gender considerations are necessary tools for conducting a professional risk assessment and planning process. Any intervention aimed at reducing risk at the individual, household or community levels should not look at women or men in isolation from each other. Instead, proposed interventions should be based on a thorough investigation of existing relations beween men and women.

4. Ask participants to refer to the case scenarios and make suggestions regarding:
   (a) interventions that would address personal gender needs of members of the community;
   (b) interventions that would address strategic gender needs.
   List suggestions on flipchart.

5. In plenary, examine how the suggestions would serve to reduce the risks of households in the communities?

6. Sum up by asking participants to say how they will use the tools of gender analysis in their work.

1. This case study is based on an exercise developed by Paulina Chiziane, Mozambique Red Cross, and facilitated in November 1994.

Notes

Resources

CASE STUDY

Case Scenario 1

Casa Village was facing serious problems of deforestation and soil erosion, caused by cutting of trees for firewood and burning of the soil. In this village, firewood was the only available resource for energy. It also served as a potential source of income.

To address these problems an aid agency discussed the situation with local authorities. They concluded that women were mainly to blame for the deforestation and erosion because they were usually the ones doing the cutting and burning.

In this village, men occupy positions of control and they are considered to be 'in charge' of women. It was therefore decided to make them responsible for changing the attitudes of women. Hence, the authorities named a large number of men to act as guardians of the forest.

The aid agency implemented projects of rural afforestation and woodland management. In order to strengthen the capacity of the villagers the agency ran training activities and introduced basic technologies in soil conservation.

The leaders of the community suggested that the beneficiaries of the project should be men; they gave the following reasons:

- *Men have the capacity to learn, because they know how to read and write.*
- *According to cultural values and local tradition, technology is men's domain.*
- *Men are the heads of households and can transmit the new learning to their mothers, wives and children.*

After a year, the agency did an evaluation and found the following:

- *Men are using the new technology in the places where they work, while women continue to work their fields in a traditional way.*
- *The relations between men and women are strained.*
- *A considerable number of men left the project.*

The agency concluded that project had not been successful and decided to look for alternative ways of improving environmental conditions.

Instructions

1. Read the case scenario, and discuss it briefly in your group.
2. Conduct a gender analysis; consider questions such as:
   - **?** What is the division of labour in productive activities?
   - **?** Who provides household income?
   - **?** Who controls the decison-making?
3. Discuss and answer the following questions:
   - 3.1. What evidence is there that a gender needs assessment had been conducted as part of the planning?
   - 3.2. What was the process of consultation and decision-making in the planning of projects?
   - 3.3. How did the agency's intervention intend to build on and support existing strengths and capacities?
   - 3.4. How did the intervention undermine existing strengths and capacities?

   Prepare a brief report back on your responses. You have 40 minutes to complete this task.

Case Scenario 2

The Forest Village is composed of repatriated people who were refugees in neighbouring countries. Upon their return they found their village devastated and although they used to grow a wide variety of crops they now depend on wild fruits and food aid. Participatory research conducted with the community showed that villagers faced many problems:

- *drought;*
- *shortage of water (the river is dry);*
- *children and adults affected by malnutrition;*
- *lack of seeds resistant to drought;*
- *lack of agricultural tools;*
- *AIDS and STD especially affecting women.*

To assist the community an international aid agency decided to target women for a number of activities:

- *training of local volunteers in primary health care;*
- *education for women on child and mother care, family planning, AIDS and STD;*
- *demonstration of food preparation based on locally available food resources;*
- *distribution of drought resistant seeds;*
- *introduction of chicken, duck and goat rearing projects.*

After a year of plentiful rains and a good harvest the agency decided to evaluate the programme and found the following results:

- *decreased hunger, especially in children, due to increase in protein intake;*
- *mothers did not apply what they had learnt about nutrition and the preparation of balanced food for their families;*
- *continued increase in the number of cases of AIDS and STD amongst women;*
- *no practice of family planning.*

The agency concluded that the project had not been successful and decided to look for alternative ways for improving villagers' conditions of life.

Instructions

1. Read the case scenario, and discuss it briefly in your group.
2. Conduct a gender analysis; consider questions such as:

   ? What is the division of labour in productive activities?

   ? Who provides household income?

   ? Who controls the decison-making?
3. Discuss and answer the following questions:

   3.1. What evidence is there that a gender needs assessment had been conducted as part of the planning?

   3.2. What was the process of consultation and decision-making in the planning of projects?

   3.3. How did the agency's intervention intend to build on and support existing strengths and capacities?

   3.4. How did the intervention undermine existing strengths and capacities?

   Prepare a brief report back on your responses. You have 40 minutes to complete this task.

How do we identify risk reduction measures?

Purpose

This activity develops participants' ability to identify possible risk reduction measures with regard to specific hazards, and to analyse how these impact on different members of a community.

Participants are reminded that it is important to consider a community's priorities and perceptions of risk when planning risk reduction. The decisions made need to be based on the community's perspective.

Procedure

This activity takes the form of a worksheet, followed by group discussion. Participants are asked to identify all the elements at risk from a hazard, and to suggest specific risk reduction measures.

Time

- 2 hours

Materials

- worksheets for each participant (see resources)
- signs with different types of hazards written on them, eg. signs saying 'drought', 'flood', 'epidemic', 'displaced persons'

Process

Introduction

1. Introduce the activity by outlining the purpose and procedure.

2. Check on participants' understanding of the meaning of the term 'risk', as "the anticipated losses (lost lives, numbers injured, property damage and disruption of economic activity) from the impact of a given hazard on a given element over a specific period of time."

A risk = hazard + vulnerability + elements at risk

3. Give a brief input in which you take a specific hazard, identify vulnerabilities, and provide an example of specific risk reduction measures. This is an example of such an input:

Let's take flood-risk as an example.

In an urban area prone to flooding some houses have been constructed in a low-lying area close to the river bank. They are made of concrete blocks and have basements or raised foundations. Other houses made of corrugated iron, cardboard and thatch have been erected in the dry river bed.

When heavy rains fall upstream and cause flooding this hazard does not affect the houses or their occupants equally. If flooding occurs the water may wash through the basements or foundations of the concrete buildings but leave the structures reasonably intact. But in the river bed the fragile dwellings are completely destroyed leaving the inhabitants destitute.

The economic vulnerability of the riverbed dwellers forced them to live in what they know is a potentially dangerous site. Their property is structurally more vulnerable than the concrete buildings. The hazard was potentially the same for both groups of inhabitants. However, it is the vulnerability (economic and structural) that has increased the risk for one group.

Two examples of risk reduction could be the following:

- *A civil engineering measure to control the river flow-rate up-stream during the rainy season.*
- *An expansion of employment opportunities, or relocation to structurally sound accommodation outside the river may reduce the vulnerability of river dwellers.*

Participant Action

1. Ask participants to get into small groups by choosing to sit under one of the 'hazard signs'. (Groups are 'full' once 4-5 participants have assembled at one sign)

2. Outline the following task (you may want to write it up on flipchart):

> -stage one: work on your own and complete the work-sheet on risk reduction;
> (20 minutes)
>
> - stage two: check your individual responses with others in the group;
> (20 minutes)
>
> - stage three: consolidate individual responses into one group response and prepare for presentation.
> (20 minutes)

3. Hand out the worksheets and briefly discuss each question, checking for understanding. Ask participants to begin working on their worksheets.

4. Monitor individual and group progress and assist where necessary.

   Ensure that groups follow the three stages of the task and prepare for presentations after approximately 1 hour.

5. Ask groups to re-convene in a plenary session, and facilitate report-backs and ensuing discussion.

Review and Discussion

1. Record the report-backs on flipchart;
   Write up the particular hazard as a heading and list the elements at risk underneath.
   — firstly, take all groups' responses to questions 2 and 3: elements at risk from the hazard;
   — secondly, take all groups' responses to question 4: risk reduction measures.

2. Ask whether the suggestions correspond to the lists of identified elements at risk. Invite participants to add to the responses.

3. Take responses to question 5: the impact of various measures.

4. Initiate a discussion around advantages and disadvantages of each suggested measure. Explore the assumptions behind each suggestion and point to the fact that risks are often perceived differently by various people.

5. Initiate a discussion around the following questions:
   ? How will the suggested activities affect the lives of women and men in a given community?
   ? How will the work involved in the risk reduction measures be distributed among women and men in the community?

6. Refer to all the records from report-backs and pose the following question:
   ? How will the risk reduction measures for one hazard reduce the risk posed by another hazard?

7. Review the activity: ask each participant to name one key thing they have learnt about risk reduction.

Resources

RISK REDUCTION WORKSHEET

PLEASE USE YOUR EXPERIENCE AND KNOWLEDGE OF TYPICAL HAZARDS AND DISASTER MANAGEMENT IN THE REGION.

1. Name the hazard you are focusing on.
2. Identify all the elements at risk from this hazard (economic, social, environmental, geographic, demographic, etc.).
3. Examine your list and prioritise elements most at risk: underline them.
4. Suggest risk reduction measures which would reduce the vulnerability of the elements most at risk.
5. Make brief notes to describe the potential impact of the risk reduction measures on various elements at risk.

How can we learn from past emergencies when planning risk reduction projects?

This exercise is more useful when participants bring maps and information on past floods with which they are familiar or which they have experienced.

Purpose

This activity aims to develop participants' ability to interpret data gathered from past flood events to forecast the likely effects of a planned risk reduction project.

Risk reduction planning requires us to link together several different disaster management components. In this activity, participants move back and forward between past and future flood events in order to examine the usefulness of a proposed flood reduction project.

Procedure

Participants are asked to project the likely impact of a future flood in a flood-prone area, based on past records, observations and experience. They then identify measures which would reduce flood impact, and consider the short and long-term effects of these projects on different at-risk groups.

Time

- 2 hours

Materials

- OHP slides (see resources)
- tracing paper, greaseproof paper or overhead transparencies
- multi-coloured pens

Process

Introduction

1. Outline the purpose and procedure of the activity.

2. Emphasise that mitigation or risk reduction planning is best when we reflect on what has occurred in past emergencies. Illustrate your point by asking:

   What would you expect to happen during this rainy season in a river valley where,over the past three years running, we have seen the same river flood? Shouldn't we expect a flood again this year?

3. Explain the following:

   This type of existing information on the impact of past floods, including their severity and extent, provides a key baseline when we plan risk reduction projects. In addition, anecdotal and other information from the community on how it coped in the past is also important for determining which interventions are likely to be the most acceptable and effective.

   We should remember that most disasters have immediate as well as longer-term effects. These include deaths, injuries, physical damage, economic and social disruption, as well as detrimental environmental effects.

4. Point out that we can use data from past floods (or other emergencies) to anticipate their future impact(s). This is most effective when we ask focused questions that provide a profile of the hazard pattern, elements at-risk and their structural/ socioeconomic vulnerability.

   — Outline a questioning sequence as an example of useful questions.

   — Explain that the purpose of this input is to help us to better understand interplay between the hazard, element at-risk and the river dwellers' vulnerability.

   — Use illustrations to demonstrate how the information generated can be used to forecast the impact of future floods.

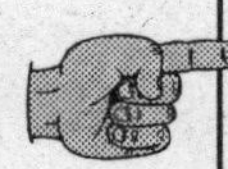

You can either use the illustrations provided (see resources): copy them on OHP slides and overlay them as you go along. Alternatively, you may want to draw your own maps, either on OHP slides or on paper.

Examples of useful questions

In the past three years a river has flooded at the beginning of the rainy season. As a result, many people lost their lives; hundreds of informal structures were washed away and crops were destroyed.

➔ **Show map 1: the image of a river bed**

Question: Which people were the worst affected by past floods in this area?
Answer: The people living right in the dry river bed.

➔ **Show map 2: houses in the river bed**

Explain that this image shows past vulnerability.
Question: Why are these families most affected by the floods?
Answer: Because this is the lowest natural point in the riverbed, and whenever there are heavy rains, the water rushes by, sweeping their houses away.

➔ **Show map 3: the same houses**

Explain that this image shows hazard behaviour + elements at-risk + vulnerability.
Question: Which families are most at-risk if a severe flood strikes again? How many families and people would be affected?
Answer: The 100 families (450 people) living right in the dry river bed.

➔ **Show map 3: the same houses**

Explain this shows anticipated future vulnerability, and estimate of population at-risk.
Question: Does this mean that the people on the river banks are not affected?
Answer: No, because about every ten years, there's a really bad flood when the river overflows its banks, but the people in the river bed are affected about every two years.

➔ **Show map 4: Water level every ten years**

Explain this shows long-term hazard behaviour + elements at risk + vulnerability.

Question: What makes some families live in the dry river bed and other live on the river banks?
Answer: The people in the river bed itself were the most recent families to arrive in this area, they could not find work, and there was no more land left on the river banks. You can tell they are poorer because most of their houses have no roofs and they are made of bits of wood and cardboard.
Explain that this illustrates socioeconomic vulnerability.

Participant Action

1. Assign participants to small working groups.

2. Ask participants to think about the impact of a past flood emergency on a community they know, and briefly exchange information and experiences on what happened.

3. Distribute tracing paper, greaseproof paper or overhead transparencies and different coloured pens to each group. Give each participant a copy of the instructions (see resources)

4. Read through the instructions and check for understanding.

5. Monitor the process of group work and assist where necessary.

Review and Discussion

1. Facilitate group report-backs. Ask reporters not to duplicate information that has already been given, but to add to previous data.

2. Check whether reportbacks include information on the following:
   - flood pattern (in terms of frequency and speed of onset, as well as duration, geographic area affected)
   - quantification of damage, losses, injuries, deaths, impact on livestock and crops.

3. Discuss the maps: do they show they flood pattern and populations most vulnerable to and affected by floods in the past?

4. Ask participants to consider different risk-reduction projects.

   Record suggestions on flipchart.

5. Examine how the different suggestions will reduce risk:
   - **?** Do they change the flood pattern, or do they alter physical or socioeconomic vulnerability?
   - **?** Estimate the costs in carrying out the projects, and the benefits (lives/property losses minimised).
   - **?** Is risk equally reduced for all community members vulnerable to flooding? Consider gender differences.

6. Initiate a discussion around issues raised:
   - **?** Why is it so difficult to measure the full impact of floods in a quantifiable way?
   - **?** What are the dangers associated with not considering recurrent flood threats when we plan development projects in flood-prone areas?
   - **?** What are the dangers associated with not considering long-term flood mitigation needs when we carry out emergency relief in flood-affected areas?

7. Ask a participant to summarise the main points of the session.

Instructions for Forecasting Activity

Please draw a series of maps:

(i) draw a map of the area discussed; include the river, location of dwellings, fields, community buildings, roads and bridges.

(ii) draw another map in a different colour, showing the area flooded in a normal rainy season.

(iii) draw a third map in another colour showing the area flooded in a "serious" flood - ie once over five-ten years.

(iv) overlay the maps on the top of each other.

Using your maps, respond to the following questions:

(1) List some of the *immediate* as well as *longer-term effects* of those floods. Who/what were most affected? How many and where?

(2) By looking at the maps and reviewing the existing information, can you *forecast the impact* of a future flood disaster with regard to

— the flood's likely behaviour (eg when and where will it strike)

— the likely elements at risk? (eg people, buildings, livestock, crops etc)?

- the vulnerability of these elements at-risk? (eg structures collapsing; crops destroyed)

(3) Draw a new map representing the flood pattern and elements at risk and as you would forecast for the next year (...years, if flooding occurs irregularly, ie only after really heavy rains).

You have 60 minutes to conclude this task.

MAP 1: Image of a riverbed

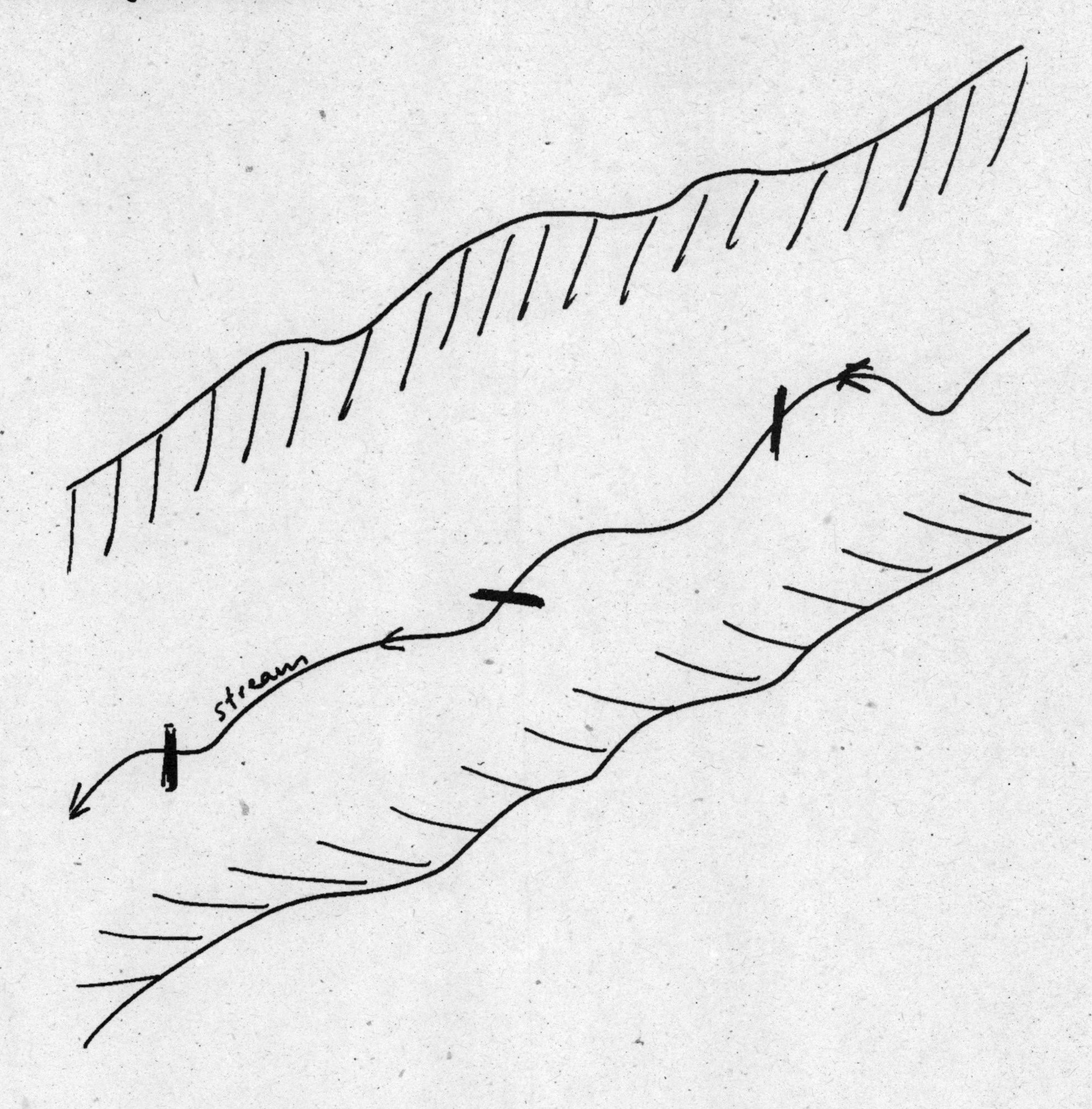

Key

river bank

stream crossing

direction of stream flow

MAP 2: Houses in the riverbed

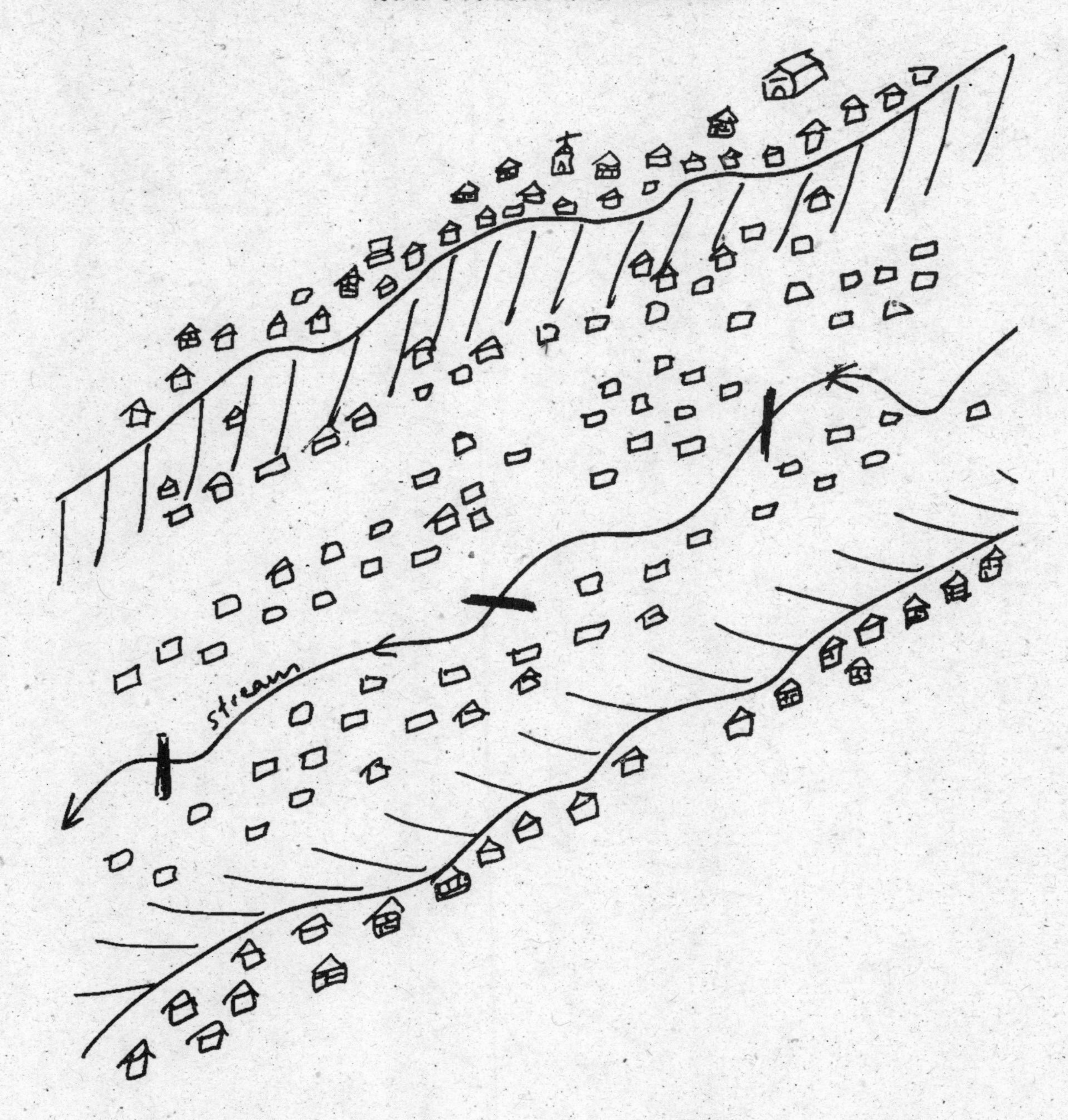

KEY

- house made of wood, cardboard
- house with corrugated roof
- house with bricks
- church
- community hall

MAP 3: Annual flood

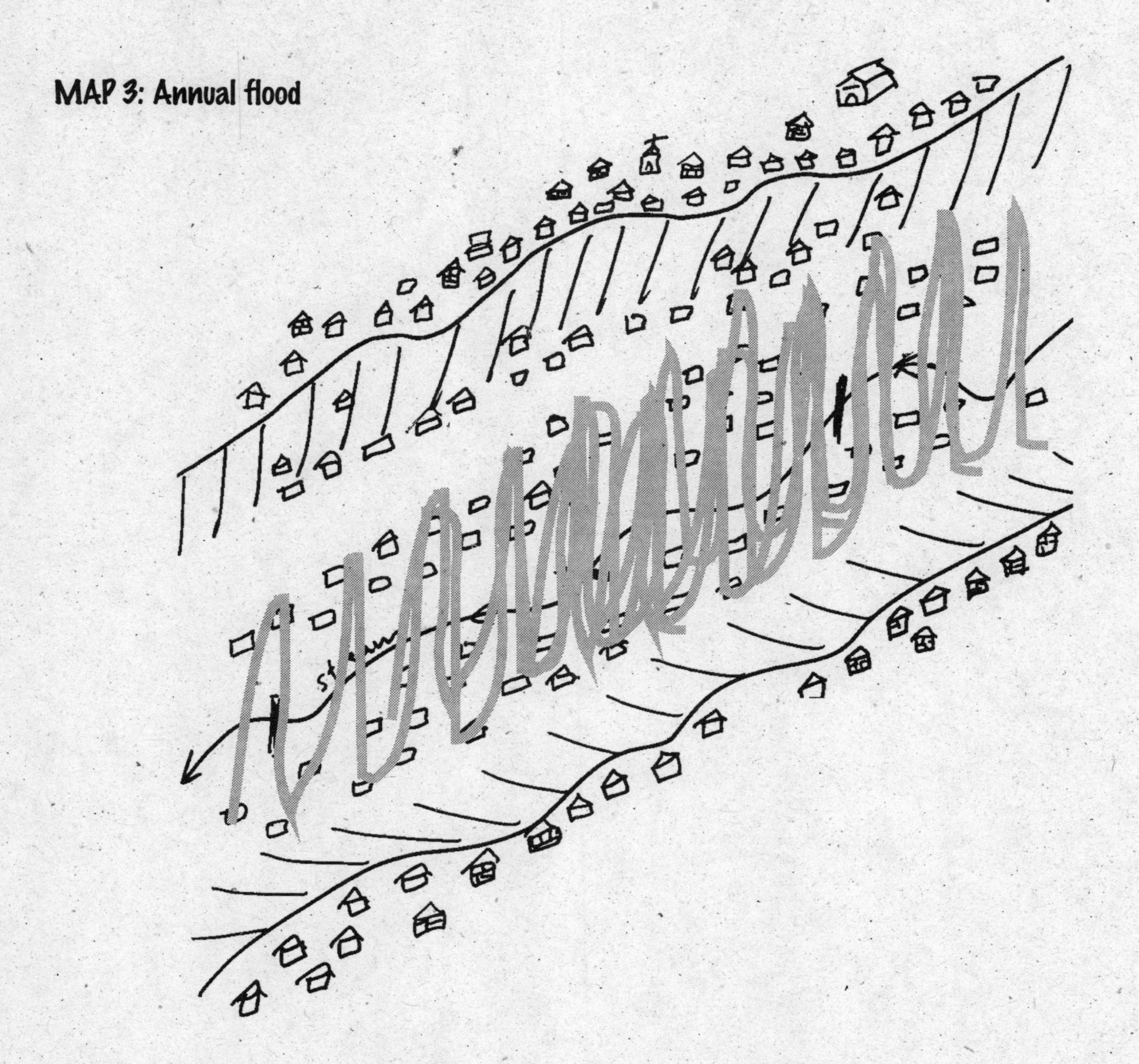

MAP 4: Flood every 10 years

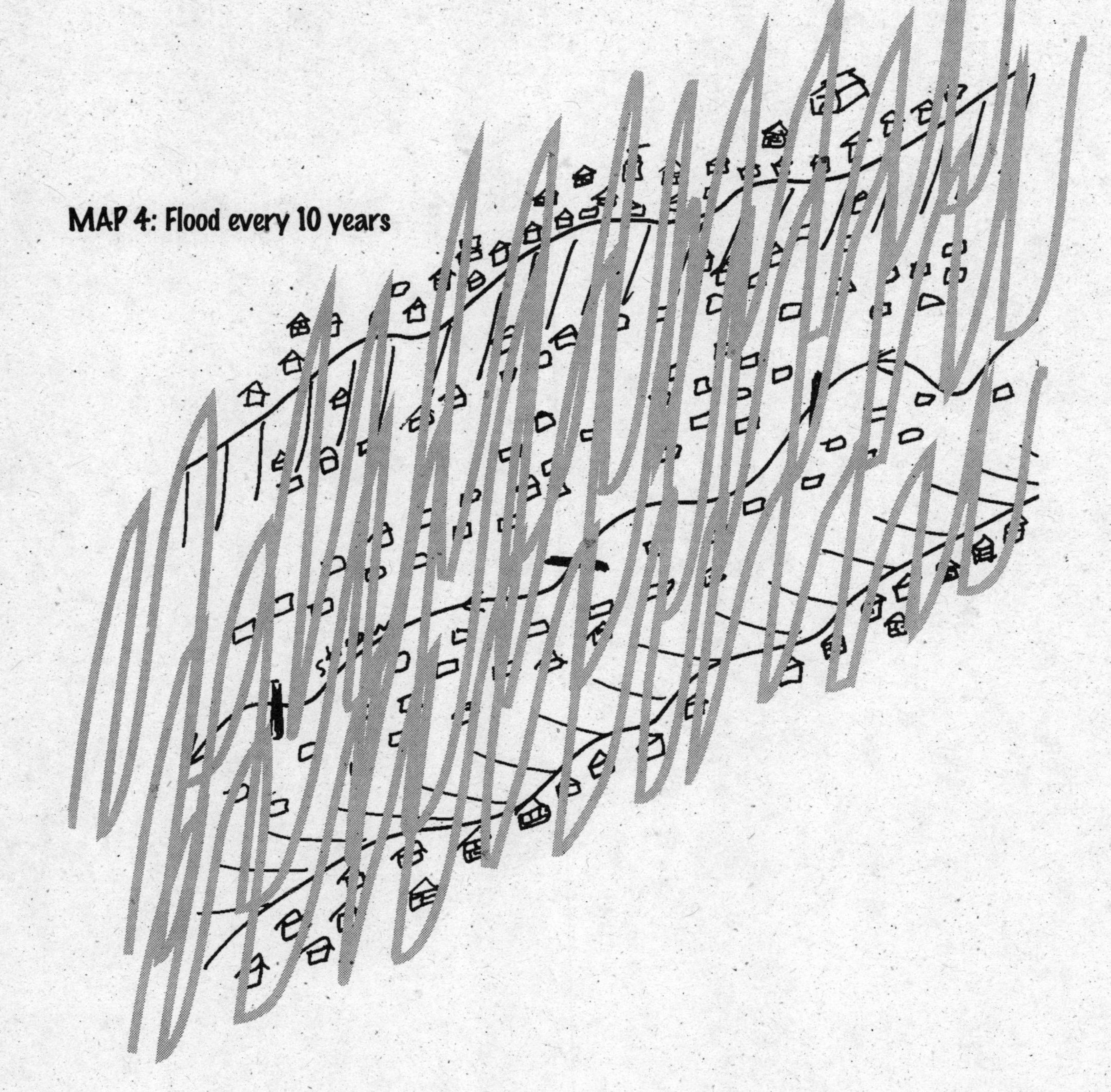

What is the role of different players in drought interventions?

Purpose

Participants will improve their knowledge of the practical work of different technical experts in drought assessment and mitigation, showing that this is a multi-sectoral effort.

Note

It is crucial that participants have received basic technical information about drought before they can conduct this activity (i.e. the difference between meteorological, hydrological and agricultural droughts).

Please see resources in Section 1, Activity 10.

Procedure

This activity has two steps. In the first participants are asked to think about the technical roles that meteorologists, hydrologists and agricultural advisors play in reducing the impact of drought. In the second, they are asked to give practical examples of drought mitigation measures which illustrate these different types of expertise and skill.

Time

- 2 hours

Materials

- briefing sheets for different working groups (see resources)

Process

Introduction

1. Outline the purpose and procedure of this activity.

2. Give the following instruction:

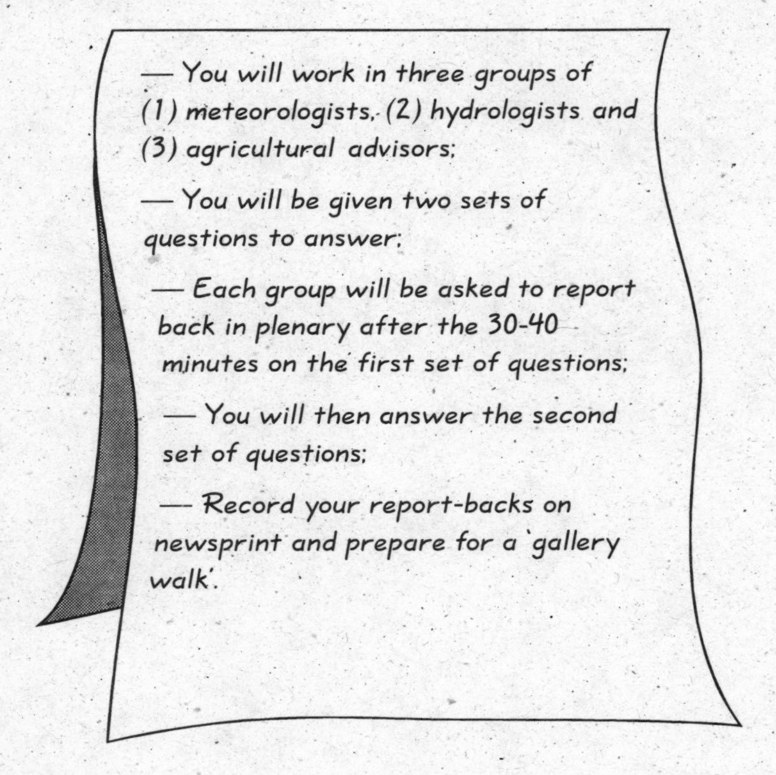

— You will work in three groups of (1) meteorologists, (2) hydrologists and (3) agricultural advisors;
— You will be given two sets of questions to answer;
— Each group will be asked to report back in plenary after the 30-40 minutes on the first set of questions;
— You will then answer the second set of questions;
— Record your report-backs on newsprint and prepare for a 'gallery walk'.

3. Check for understanding of terminology by asking participants to define 'meteorologist', 'hydrologist' and 'agricultural advisor'.

Participant Action

1. Ask participants to get into three groups and hand out briefing sheets.

2. Monitor progress and assist where necessary.

Review and Discussion

1. In plenary, ask groups to report back their responses to the first set of questions. Encourage questions of clarification, and discussion.

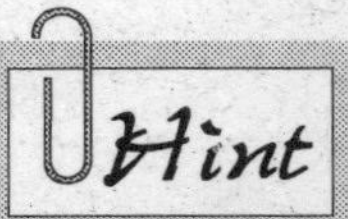

Issues you might want to touch on include the following:

— Different players have different roles with regard to drought prediction, warning, prevention, mitigation and response. For example, meteorologists play key roles in weather forecasting, prediction and warning, but are not operational players in actual mitigation. Hydrologists and agriculturalists on the other hand also have forecasting and prediction roles, but they do have operational capacities.

— Sometimes information on rainfall, water supply and food security is not widely disseminated beyond technical specialists. To make a real impact this data must be available to practitioners as well as scientists.

— Vulnerability to drought is not the same from one community, to another. Therefore scientists also need field information from practitioners, like NGO workers, on which communities are most vulnerable. This allows them to determine where the drought's impact is greatest.

— These principles also apply to other hazards besides drought.

Participant Action

1. Request participants to return to their groups and address the second set of questions. Ask them to write their responses on flipchart. Allow approximately 20 minutes for group work.

Review and Discussion

1. Ask groups to display their report-backs and invite participants to go on a 'gallery walk'.
2. In plenary, ask for questions of clarification.
3. Highlight gaps and strengths in the report-backs.

   Point out those specialised meteorological, hydrological and agricultural measures known to make a difference. This illustrates the need for NGOs to draw on these specialised skills, and not to work in isolation.

Notes

Resources

ROLE PLAY

Individual briefing sheets for role players

Hydrologists

First Set of Questions

You are all hydrologists working in Southern Africa.

Please consider a drought you are most familiar with, and respond to the following questions:

(1) What role did you play with respect to

(i) mitigating / preventing the drought's worst consequences?

(ii) predicting and warning about the drought?

(iii) responding and recovery?

(2) Who used your findings? Who else would you like to use your findings?

(3) How could NGOs help you to improve the accuracy of your data?

Second Set of Questions

(1) From your experience:

(i) Give examples of water storage / supply systems (e.g. dams, wells etc.) that withstood the drought's impact, and those that didn't. In what way were they different?

(ii) What traditional / other methods increase or decrease the moisture retention of sandy soils?

(2) Consider hazard reduction measures: in an area that is repeatedly drought-stricken, what could you as hydrologists do to reduce the severity of future droughts?

Meteorologists

First Set of Questions

You are all meteorologists working in Southern Africa.

Please consider a drought you are most familiar with, and respond to the following questions:

(1) What role did you play with respect to

(i) mitigating / preventing the drought's worst consequence?

(ii) predicting and warning about the drought?

(2) Who used your findings? Who else would you like to use your findings?

(3) How could NGOs help you to improve the accuracy of your data?

Second Set of Questions

(1) (i) Give examples of other meteorological droughts from your own experience.

(ii) Outline a number of possible causes of rainfall deficit that may have lead to the drought.

(2) Consider risk reduction measures: in an area that is repeatedly drought-stricken, what could you as meteorologists do to reduce the impact of metereological droughts on water systems and local agriculture?

Agricultural Advisors

First Set of Questions

You are all agricultural advisors working in Southern Africa.

Please consider a drought you are most familiar with and respond to the following questions:

(1) What role did you play with respect to

(i) mitigating/ preventing the drought's worst consequence?

(ii) predicting and warning about the drought?

(iii) responding and recovery?

(2) Who used your findings? Who else would you like to use your findings?

(3) How could NGOs help you to improve the accuracy of your data?

Second Set of Questions

(1) (i) From your own experience, give examples of the drought's agricultural impact.

(ii) Outline the factors that increased the vulnerability of certain population groups.

(2) Consider measures to reduce the agricultural vulnerability of those groups: in an area that is repeatedly drought-stricken, what could you as agricultural advisors do to improve food and livelihood security?

How do we prepare public awareness campaigns?

Purpose

Participants improve their understanding of how to design an effective public awareness campaign as a tool for information-giving aimed at risk reduction.

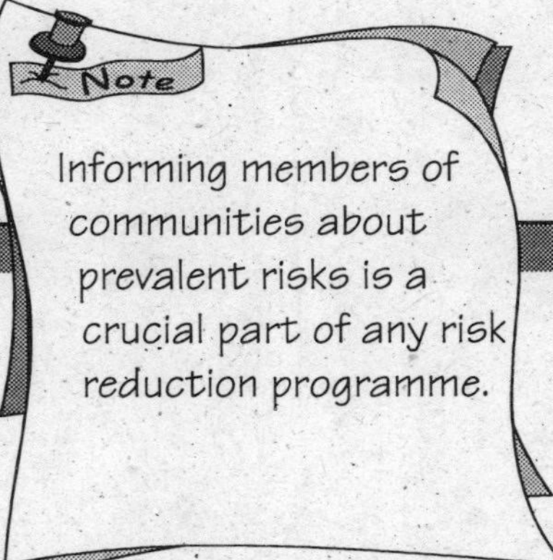

Note
Informing members of communities about prevalent risks is a crucial part of any risk reduction programme.

Procedure

In this activity participants will design plans for a public awareness campaign, and examine them critically for their effectiveness.

Time

- 2 hours

Materials

- tape recorder (optional)

Process

Introduction

1. Introduce the activity by outlining the purpose and procedure.

2. Ask participants whether any of them have ever been involved in a public awareness campaign. Ask those with experience to briefly talk about what happened.

3. Point out that a public awareness campaign is similar to any other form of communication: to be true communication the message sent requires a *response* from a *receiver*. If the message sent out finds no echo, it has been sent in vain. For example: if a public awareness campaign aims at a decrease in the risk of infection of cholera, the response should be a change in the behaviour of the public, which in turn results in a decrease of infection.

4. Point out that the purpose of a planned campaign should be specific and clear, and the response should at best be observable / measurable in some way.

5. Give a brief input on the five basic planning questions:

   - **?** Who? (is the target audience)
   - **?** What? (is the aim of the campaign? What do you want to achieve?)
   - **?** How? (will you go about raising the awareness? What tools will you use?)
   - **?** When? (will you launch the campaign? Is there a special occasion to raise the profile of the campaign?)
   - **?** Where? (will you effect the campaign? Where will you perform the plays / hang up the posters / distribute pamphlets, etc.)

Participant Action

1. Introduce the task by explaining that this activity challenges participants' creativity and imagination - two of the most important elements for problem-solving.

2. Give the following instruction:

Topic: THE RISING RATE OF HIV

You are asked to design a public awareness campaign:

— You will work in three groups:
  - group 1 will use a play;
  - group 2 will use the radio;
  - group 3 will use a poster campaign.

— Your target group are urban informal settlements / squatter camps

— You will be asked to perform / present your campaign plans: be inventive!

You have 45 minutes to prepare

3. Divide participants into three groups and advise them to begin their design by answering the five basic planning questions.

4. Monitor the process and assist where necessary.

5. Ask groups to present their campaigns.

Presentations and Discussion

1. Unpack each presentation by asking questions such as:
   - **?** What was good about it? What did you like?
   - **?** What were its limitations?
   - **?** What did you learn about the risk of HIV? (Why / how did you learn it?)
   - **?** Was it clear what change in behaviour or response would reduce the risk of HIV infection? Why / why not?

2 Discuss some of the difficulties that are likely to arise in the presentations, such as:
   - — lack of clear focus
   - — attempt to transmit too much information
   - — ambiguity of message
   - — unattractive medium /design
   - — tension between your agenda and recipients' need

3. Establish criteria for designing public awareness campaigns aimed at reducing risk. List criteria on flipchart.

4. Sum up by asking participants to state what advice they would give to someone embarking on a public awareness campaign that will be part of a risk reduction programme.

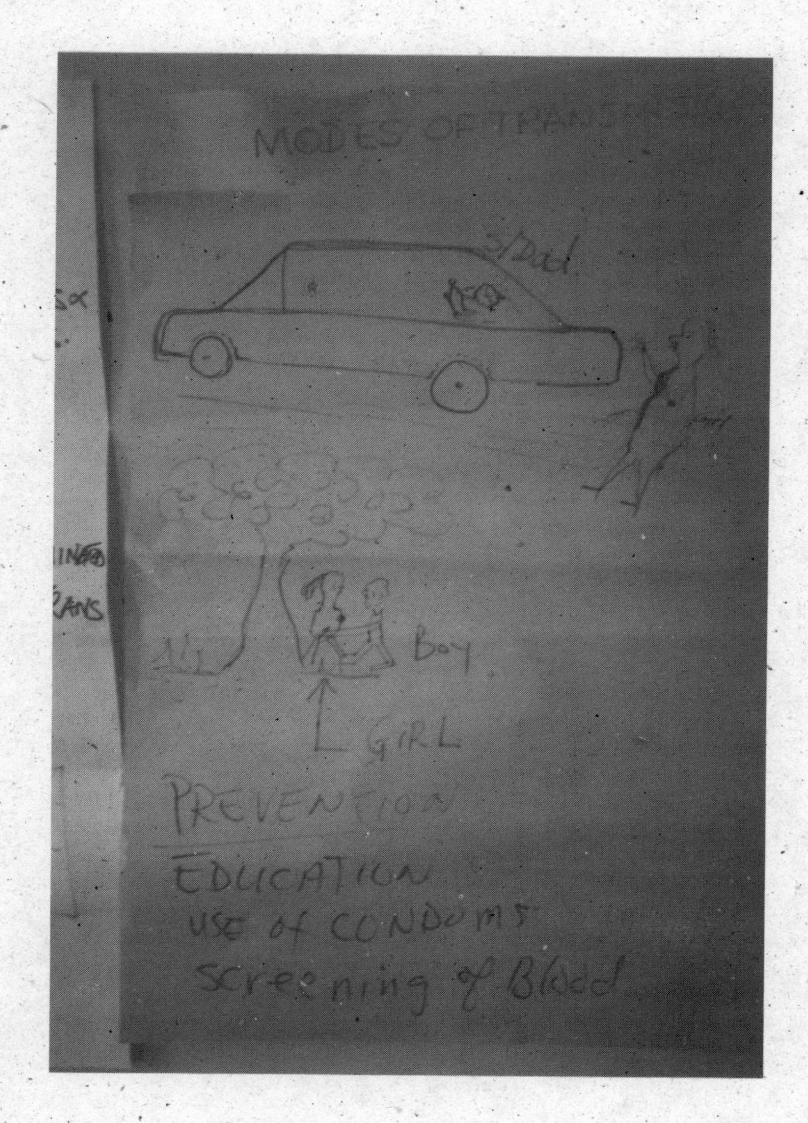

What is the role of development workers in risk reduction?

Purpose

This discussion-based activity asks participants to make decisions about aspects of development work.

Note

This activity requires differentiated and critical thinking and challenges participants to expand their understanding of development issues.

Procedure

Participants are confronted with statements about development issues and their role as development workers. Discussing each statement in a small group, participants have to reach consensus about whether they agree or disagree with the statements.

Time

- 1 hour

Materials

- one set of cut-up statements about development for each group (see resources)
- 3 'postboxes' (such as pockets made by open envelopes) with the labels 'agree', 'disagree', and 'undecided' for each group

Process

Introduction

1. Outline the purpose and procedure of the activity.

2. Ask participants to get into small groups of no more than 5 members and ask them to sit in a semi-circle in front of a wall.

3. For each group, stick the 3 labelled envelopes onto the wall; point out that these are 'postboxes'.

Participant Action

1. Give each group a set of statement cards stacked in a random pile. Ask them to put the cards in the centre of the semi-circle where everyone can reach them.

2. Give the following instruction:

> - participants will take turns in picking up a card and reading it out aloud to the group;
>
> - the group will discuss whether they agree or disagree with the statement;
>
> - when the group has reached a consensus decision the statement is 'posted' into the appropriate 'agree' or 'disagree' envelope;
>
> - if the group cannot reach consensus the statement is placed into the 'undecided' envelope;
>
> - when all statements have been 'posted' the group can return to the 'undecided' box and re-open discussion around the statement.

3. Set a time limit (approx. 30 minutes) and point out that thorough discussion is more important than quick decision-making.

4. Monitor progress and assist by clarifying, if necessary.

Review and Discussion

1. When groups have completed their discussion or when the time allocated has elapsed initiate a plenary discussion.
   Include questions such as the following:
   - **?** Which statements were / are in the 'undecided' box and what made a decision difficult? What information was needed to reach consensus?
   - **?** Which statements raised critical issues around development and / or with regards to the role of development workers?
   - **?** Which statements offered a new / different perspective to participants?

2. Ask participants to apply some of the issues raised to their work in risk reduction: how will they use in the field what they have learnt in this discussion?

3. Sum up key issues from this activity.

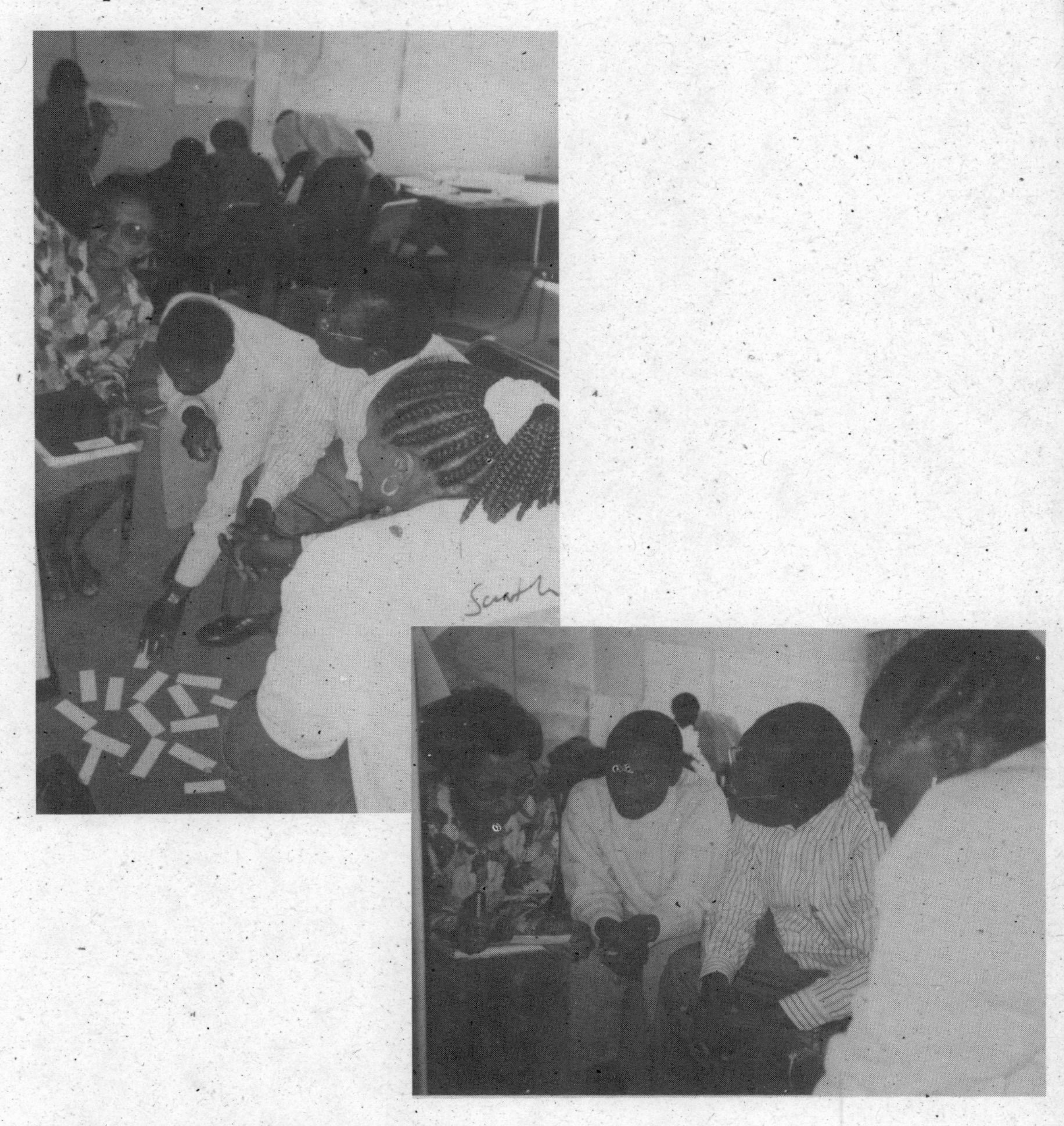

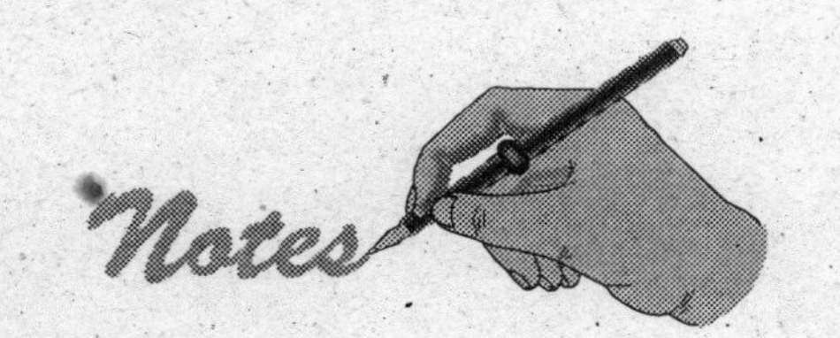
Notes

DISCUSSION STATEMENTS

Development workers aim to make people self-reliant.

A development worker should concentrate primarily on projects that lead to food security.

Women are more open to change than men.

Agrarian reform is a necessary pre-condition for the elimination of hunger.

Development workers should engage people to 'think the unthinkable' and imagine what is possible

Development workers must talk with people, not to them.

Poor people have a poor grasp of the causes of their economic and social problems.

Development workers gain the confidence of communities by demonstrating their skills and knowledge.

The role of the development worker is to search with the community for the causes of problems.

Most rural communities are open to collective ownership and management of economic activities.

Self-reliance must lead to complete self-sufficiency.

Development workers can work from the assumption that producers and traders have common interests.

True development is based on a series of analysis - action - reflection carried out by the poor.

Development workers should facilitate discussions until community members reach consensus decisions.

There cannot be development without technology.

Food aid often results in changed food habits and this causes new dependencies.

Low self-esteem is a characteristic of poor people.

Crops that give 'high yields' are more useful than crops that give 'stable yields'.

Useful References

Abbat, F.R. (1992). *Teaching for better learning: a guide for teachers of primary health care staff.* Geneva: World Health Organisation.

Adirondack, S. (1992). *Just about managing: effective management for voluntary organisations and community groups.* London: London Voluntary Service Council (LVSC).

Anderson, M.B. and Woodrow, P. (1989). *Rising from the ashes: development strategies in times of disaster.* Westview Press, Boulder, Colorado INC & UNESCO.

Anderson, M.B. and Woodrow, P.J. (1990). *Disaster and development workshops: a manual for training in capacities and vulnerabilities analysis.* Harvard University: International Relief/ Development Project, Harvard

Anderson, M.B. and Woodrow, P.J. (n.d.). Reducing vulnerability to drought and famine: developmental approaches to relief. *Disasters*, Vol 15 No 15.

Blaikie, P. et. al. (1994). *At risk: natural hazards, people's vulnerability and disasters.* New York: Routledge.

Burkey, S (1993). *People first: a guide to self-reliant participatory rural development. London:* ZED Books Ltd.

Buzzard, S. and Edgecomb, E. (1987). *Monitoring and evaluating small business projects: a step by step guide.* Small Enterprise Evaluation Project (SEEP) of PACT.

Cammack, J. (1992). *Basic accounting for small groups.* Oxford: Oxfam.

Carr, M. (ed). (1991). *Women and food security: the experience of the SADCC countries.* London: IT Publications.

Carter, W.N. (1991). *Disaster management: a disaster manager's handbook.* Manila: Asian Development Bank.

Chambers, R. (1994). Participatory Rural Appraisal (PRA): challenges, potentials and paradigm. *World Development*, 22 (10): 1437-1454. Great Britain: Elsevier Science Ltd.

Chambers, R. (1994). Participatory Rural Appraisal (PRA): analysis of experience. *World Development*, 22 (9): 1253-1268. Great Britain: Elsevier Science LTD.

Chambers, R. (1994). The origins and practice of Participatory Rural Appraisal. *World Development*, 22 (7): 953-969. Great Britain: Elsevier Science Ltd.

Chowdhury, R. et. al. (1992). The Bangladesh cyclone of 1991: why so many died. *Disasters*, 17 (4).

Cross, N. and Barker, R. (c.1992). *At the desert's edge: oral histories from the Sahel.* London: Panos Publications Ltd.

Elson, D. (1991). Structural adjustment: its effects on women. In: *Changing Perceptions. Writings on Gender and Development.* Oxford: Oxfam.

Fernand, V. (1989). *Manual of practical management: organisation administration communication, vol I.* Development Innovations and Networks (IRED).

Fernand, V. (1989). *Manual of practical management: financial management, vol II.* Development Innovations and Networks (IRED).

Feuerstein, M.T. (1986). *Partners in evaluation: evaluating development and community programmes with participants.* London: Macmillan Publishers.

Grandin, B.E. (1988). *Wealth ranking in smaller holder communities: a field manual.* UK: Intermediate Technology Publication, Ltd.

Greet, P. (1994). Making good policy into good practice. *Focus on Gender* 2 (1).

Henderson, P (1989). *The effective trainer series: promoting active learning.* Cambridge: National Extension College.

International Decade for Natural Disaster Reduction (1991). First Session of the Scientific and Technical Committee. *STOP Disasters*, No 1.

International Women's Tribune Centre (IWTC). (1990). *Women and Water: a collection of IWTC newsletters on issues, activities and resources in the area of women, water and sanitation needs.* New York: IWTC.

Kindervatter, S. (1987). *Women working together: for personal, economic and community development.* Washington, DC: OEF International.

Kindervatter, S. (1991). *Appropriate business skills for third world women. Doing a feasibility study: training activities for starting reviewing a small business.* Washington DC: OEF International and UNIFEM

Maskrey, A. (1989). *Disaster mitigation - a community based approach: approaches to mitigation.* Oxford: Oxfam.

McPherson, C. et al. (1989). *Disasters in the classroom: teaching about disasters in the third world.* Oxford: Oxfam Education Department.

Meena, R. (ed). (1992). *Gender in Southern Africa: conceptual and theoretical issues.* Harare: SAPES Books.

Moser, C. (1989). Gender planning in the third world meeting practical and strategic gender needs. *World Development* 17 (11). Great Britain.

Myers, M. (1994). Women and children first: introducing a gender strategy into disaster on preparedness. *Focus of Gender* 2 (1).

Nichols, P. (1991). *Social survey methods: a field guide for development workers.* Oxford: Oxfam.

Pushpanath, K (1994). The Zambian drought experience: interview. *Focus on Gender* 2 (1).

RRA Notes 19. (1994). *Special issue of training.* London: IIED.

Sandhu, R. and Sandler, J. (1986). *The tech and tools book: a guide to technologies women are using worldwide.* Great Britain: International Women's Tribune Centre and Intermediate Technology Publications.

SARDC and IUCN (1994). *State of the environment in Southern Africa.* Zimbabwe: SARDC, IUCN, SADC.

Scrimshaw, N.S. and Gleason, G.R. (1992). *Rapid assessment procedures: qualitative methodologies for planning and evaluation of health related programmes.* Boston: INFDC.

Shariff, H. (1990). *Traditional ways of helping each other.* Geneva: Institut Henry-Dunant.

Srinivasan, L. (1992). *Options for educators: a monograph for decision makers on alternative participatory strategies.* New York: PACT/CDS, Inc.

Training with UNHCR. (1992). *Coping with stress in crisis situations.* Geneva: UNHCR Training Service.

UNDRO. (1990). *Mitigating natural disasters, phenomena, effects and options: a manual for policy makers and planners.* Geneva: UN Publications.

UNDP, DHA Disaster Management Training Programme, in association with the University of Wisconsin Disaster Management Centre. (c. 1990). *Introduction to hazards.* Wisconsin: Editorial Services, incl. Design, educational components and formatting by Interworks.

Vella, J. (1989). *Learning to teach: training of trainers for community development.* Washington, DC: OEF International.

von Kotze, A. (1992). *Clapslap juggle and zoom: energising games for adult educators.* Durban: Centre for Adult Education, University of Natal.

Werner, D. and Bower B. (1982). *Helping health workers learn: a book of methods, aids, and ideas for instructors at the village level.* Palo Alto: Hesperian Foundation

World Health Organisation. (1988). *Education for health. A manual on health education in primary health care.* Switzerland: World Health Organisation.